Kinetics of Soil Chemical Processes

Kinetics of Soil Chemical Processes

DONALD L. SPARKS

College of Agricultural Sciences
Department of Plant Science
University of Delaware
Newark, Delaware

ACADEMIC PRESS, INC.
Harcourt Brace Jovanovich, Publishers
San Diego New York Berkeley Boston
London Sydney Tokyo Toronto

ACADEMIC PRESS, INC.
San Diego, California 92101

United Kingdom Edition published by
ACADEMIC PRESS, INC. (LONDON) LTD.
24-28 Oval Road, London NW1 7DX

Library of Congress Cataloging-in-Publication Data

Sparks, Donald L., Ph. D.
Kinetics of soil chemical processes / Donald L. Sparks.
p. cm.
Bibliography: p.
Includes index.
ISBN 0-12-656440-X (alk. paper)
1. Soil physical chemistry. 2. Chemical reaction, Rate of.
I. Title.
S592.53.S67 1988
631.4′1—dc19 88-14656
CIP

PRINTED IN THE UNITED STATES OF AMERICA
89 90 91 92 9 8 7 6 5 4 3 2 1

To my wife, Joy, and my late mother, Christine McKenzie Sparks,
with love and respect

Contents

4 Kinetics and Mechanisms of Rapid Reactions on Soil Constituents Using Relaxation Methods

5 Ion Exchange Kinetics on Soils and Soil Constituents

6 Kinetics of Pesticide and Organic Pollutant Reactions

7 Rates of Chemical Weathering

8 Redox Kinetics

9 Kinetic Modeling of Inorganic and Organic Reactions in Soils

Preface

Kinetics of reactions in soil and aquatic environments is a topic that is of extreme importance and interest. Most of the chemical processes that occur in these systems are dynamic, and a knowledge of the mechanisms and kinetics of these reactions is fundamental. Moreover, to properly understand the fate of applied fertilizers, pesticides, and organic pollutants in soils with time, and to thus improve nutrient availability and the quality of our groundwater, one must study kinetics.

Before the publication of this book, no comprehensive treatment of these concepts existed. This book fully addresses the above needs. It should be useful to students and professionals in soil science, geochemistry, environmental engineering, and geology. Chapter 1 introduces the topic of kinetics of soil chemical processes, with particular emphasis on a historical perspective. Chapter 2 is a comprehensive treatment of the application of chemical kinetics to soil constituents, including discussions of rate laws and mechanisms, types of kinetic equations, and transition state theory.

Perhaps the most important aspect of any kinetic study is the methodology one uses. Chapter 3 discusses different kinetic methodologies, their advantages and disadvantages, and how one can interpret the data gained from them.

Many reactions on soils and soil constituents are very rapid, occurring on millisecond time scales. Chapter 4 discusses these types of reactions and how they can be measured using pressure-jump (p-jump) relaxation, stopped flow, and electric field pulse techniques. One of the salient reasons for studying the kinetics of reactions on soils is to gather information on rate-limiting steps. This is discussed in Chapter 5, along with other aspects of ion exchange kinetics. Currently, one of the major research areas in soil and environmental chemistry is the interactions of pesticides and organic pollutants with soils. Chapter 6 discusses the rates and mechanisms of these interactions. Chapter 7 deals with kinetics of chemical weathering, including dissolution rates and mechanisms of feldspar, oxide, and ferromagnesian minerals. Chapter 8 discusses redox kinetics, including reductive dissolution of oxides and oxidation of inorganic ions in soils and sediments. Perhaps one of the most exciting and challenging approaches to integrating and synthesizing a large body of knowledge on soil elemental transformations and transport is mathematical modeling and computer simulation. Modeling reactions kinetically is certainly a useful approach for predicting the fate of applied fertilizers, pesticides, and toxic organics with time in soil and aquatic environments. A discussion of kinetic modeling is given in Chapter 9.

I would like to express my gratitude to a number of people who encouraged and assisted me in the writing of this book. I am indebted to Professor Garrison Sposito, who first encouraged me to write a book on kinetics of soil chemical processes. Much of the initial planning for this book was accomplished while I spent a sabbatical leave at the University of California, Riverside. My interactions there with Professors Sposito, the late F. T. Bingham, and P. F. Pratt were particularly stimulating. I am also deeply indebted to the University of Delaware for providing me such a wonderful environment in which to conduct research and to teach. Additionally, I owe much appreciation to the group of fine graduate students and postdoctoral associates with whom I have worked at the University of Delaware over the past nine years. Special thanks are extended to Ted Carski, Richard Ogwada, Phil Jardine, Mark Seyfried, Cris Schulthess, Mark Noll, Asher Bar-Tal, Clare Evans, Harris Martin, Steve Grant, Nanak Pasricha, Matt Eick, Z. Z. Zhang, Jerry Hendricks, Maria Sadusky, Pengchu Zhang, and Chip Toner for their stimulating research and helpful discussions. I owe special gratitude to Professors S. Kuo, S. Feagley, T. Yasunaga, and Garrison Sposito for their careful and thoughtful reviews of the manuscript. I am also especially indebted to my secretary, Muriel Ryder, for typing the manuscript and for her splendid assistance during my career at the University of Delaware. I also thank Keith Heckert for the fine art work in this book. Special thanks are also extended to the staff of Academic Press for encouragement and helpfulness. Of course, none of this could have been accomplished without the love, support, and encouragement of my wife, Joy, and that of my parents. To them, I shall be eternally grateful.

Index of Principal Symbols

A	frequency factor; cross-sectional area of reactor
a,b,y	stoichiometric coefficients
$(AB)^*$	activated complex
α, β, etc.	partial reaction order
$\alpha_{A/B}$	separation factor
b	y-intercept
$\underline{C}$	concentration of sorptive
$\bar{C}$	total counterion concentration on ion exchanger
C_{eff}	effluent concentration
C_{eq}	equilibrium concentration
C_f	final concentration
$\underline{<\bar{C}_i>}$	average concentration of ion i on ion exchanger
$\bar{C}_i$	concentration of counterion i on the ion exchanger
C_{in}	influent concentration
C' (r)	compound concentration free in the pore fluid changing with radial distance (r)
C_0	initial concentration
C_p	concentration of particle
C_s	concentration in bulk solution
C_T	total concentration of sorptive or desorptive
c_j	molar concentration of ion j
σ	conductance
$\underline{D}$	diffusion coefficient
$\bar{D}$	interdiffusion coefficient
D_c	dispersion coefficient
D_{eff}	effective particle diffusion
D_f	film diffusion coefficient
D_g	distribution coefficient
D_m	pore fluid diffusivity
d	depth or distance
Θ_H	degree of surface protonation
ψ	electrical potential
E	energy of activation
E_a	energy of activation for adsorption
E_d	energy of activation for desorption
E_1	energy of activation for forward reaction
E_{-1}	energy of activation for backward reaction
ξ	extent of reaction
F	Faraday constant
ΔG^0_{ex}	standard free energy of exchange

$\Delta G^{\ddagger}$	Gibbs energy of activation
$\Delta G_{\circ}^{\ddagger}$	standard Gibbs energy of activation
$\Delta G_1^{\ddagger}$	Gibbs energy of activation for forward reaction
δ	film thickness
ΔH_{ex}°	standard enthalpy of exchange
$\Delta H^{\circ\ddagger}$	standard enthalpy of activation
$\Delta H_1^{\ddagger}$	enthalpy of activation for forward reaction
h	Planck's constant
J	flow rate or ion flux
K	rate of product formation
K_{eq}	equilibrium constant
K_{ex}	exchange equilibrium constant
K_L	pseudo-Langmuir constant
$K_{0,ap}$	apparent equilibrium constant
K_p	equilibrium partition coefficient
K_t	transmittance coefficient
$K^{\ddagger}$	pseudothermodynamic equilibrium constant of activated complex
k	rate constant
k_a	adsorption rate coefficient
k_a'	apparent adsorption rate coefficient
k_B	Boltzmann constant
k_d	desorption rate coefficient
k_d'	apparent desorption rate coefficient
k_F	Freundlich coefficient
k_q	sorption rate coefficient
k_1	forward rate constant
k_{-1}	backward rate constant
$k_1^{\ddagger}$	rate of formation of activated complex
$k_{-1}^{\ddagger}$	rate of decomposition of activated complex
ℓ	length
$(M\text{-}q)$	surface unsaturation
m	mass or slope of line
m_j	molar concentration of ion j
n	overall reaction order
η	error term in two-constant rate equation
p	pressure
p_s	porosity of sorbent
Q_t	quantity of diffusing substance leaving cylinder at time t
$Q\alpha$	quantity of diffusing substance leaving cylinder at infinite time
Γ, Γ^{∞}, Γ_t	equilibrium, saturated, and total amount of sorbed ion or molecule
q	amount or quantity of ion or molecule sorbed
R	Universal gas constant
r_c	reaction capacity term
r_d	distance from particle center
r_o	radius of particle
ρ_B	bulk density of sorbent or density of solution
ρ_s	specific gravity
$S(r)$	local total volumetric concentration in porous sorbent
ΔS_{ex}°	standard entropy of exchange
$\Delta S^{0\ddagger}$	standard entropy for activation
$\Delta S_1^{\ddagger}$	entropy of activation for forward reaction

T	absolute temperature
t	time
t_o	integration constant
tor	intraaggregate porosity or tortuosity
$t_{1/2}$	reaction half-life
τ	relaxation time
$U(t)$	fractional attainment of equilibrium
u_j	electrical mobility
u_{pw}	pore water velocity
V	volume
ΔV	standard molar volume change of reaction
ν	reaction rate
Θ	volumetric water content
X, Y	Elovich equation constants
$\bar{y}$	partial molar activity coefficient on the exchanger phase
Z	reciprocal of rate or $(dq/dt)^{-1}$
z	valence

1

Introduction

Kinetics of soil chemical processes is one of the most important, controversial, challenging, enigmatic, and exciting areas in soil and environmental chemistry. Every year the interest and research activity in this area increase. Yet there are many unanswered questions. Certainly there are many avenues of research dealing with reaction rates and mechanisms on soils and soil constituents that need pursuing.

Since the inception of soil chemistry, much attention has been given to equilibrium processes involving soils, humic materials, clay minerals, and sediments. Without question, the results of these studies have proved enlightening and beneficial. However, they have not provided information on the kinetics of these reactions. Moreover, equilibrium studies are often not applicable to field conditions, since soils and sediments are nearly always at disequilibrium with respect to ion transformations and organic molecule interactions (Sparks, 1987a). To properly and completely understand the dynamic interactions of pesticides, organic pollutants, sludges, wastes, and fertilizers with soils and sediments, a knowledge of the kinetics is fundamental.

The first studies on kinetics of soil chemical processes appeared in the early 1850s in the remarkable ion exchange research of J. Thomas Way (1850). He found that the rate of NH_4^+–Ca^{2+} exchange on a British soil was extremely rapid, almost instantaneous. Similar conclusions were reached later by Gedroiz (1914) in Russia and Hissink (1924) in the Netherlands. The results of these workers went unquestioned for many years. In 1947, a remarkable landmark paper on ion exchange kinetics appeared by Boyd *et al.* These investigators, working in conjunction with the Manhattan Project of World War II, clearly established that ion exchange was diffusion-controlled. Additionally, they were the first to elucidate mechanisms and rate-determining steps for ion exchange phenomena. As we shall see later (Chapter 5), the impact their work has had on kinetic investigations, particularly involving cation and anion exchange research, is truly remarkable.

Kelley (1948) published his beautiful book "Cation Exchange in Soils" and astutely and accurately hypothesized that while ion exchange rates may be very rapid on kaolinite and smectite, reactions on other 2 : 1 clay minerals such as vermiculite and mica could be quite slow. He also reasoned that exchange rates were affected by ion type and size. For example, he believed that exchange of K^+ and NH_4^+ on micas and vermiculites was not instantaneous owing to these clays' stereochemistry and geometry and to the size characteristics of the ions. Unfortunately, Kelley's ingenious observations on kinetics of ion exchange went relatively unnoticed.

From the period of the 1920s to the late 1950s little work appeared in the soil and environmental chemistry literature on kinetics. One can only speculate as to the reasons for this. Perhaps many researchers accepted without question the early work of Way (1850), Gedroiz (1914), and Hissink (1924) that kinetic reactions in soils were instantaneous.

Another impediment to studying kinetic processes in soils in the early days and even now is methodology-related (Chapter 3). With traditional batch techniques, where centrifugation is employed to obtain a clarified supernatant, reaction rates less than ~5 min cannot be observed. We now know that many reactions on soils and sediments are exceedingly rapid, occurring on millisecond and even microsecond time scales.

In the late 1950s and early 1960s a series of papers appeared by Mortland (1958), Mortland and Ellis (1959), and Scott and Reed (1962) on the dynamics of potassium release from biotite and vermiculite. These researchers were among the first soil chemists to apply chemical kinetic principles to soil constituents and to elucidate rate-limiting steps. About this time, Keay and Wild (1961), investigating kinetics of ion exchange on vermiculite concluded that particle diffusion (PD) was rate-limiting, calculated energies of activation (E) for the exchange processes, determined diffusion coefficients for vermiculites ($\bar{D}$) (Chapter 5) and were among the first to use kinetics to obtain pseudothermodynamic parameters using Eyring's reaction rate theory (Chapter 2).

Also, in the late 1950s and 1960s some particularly seminal papers on ion exchange kinetics appeared by Helfferich (1962b, 1963, 1965) that are classics in the field. In this research it was definitively shown that the rate-limiting steps in ion exchange phenomena were film diffusion (FD) and/or particle diffusion (PD). Additionally, the Nernst–Planck theories were explored and applied to an array of adsorbents (Chapter 5).

The application of chemical kinetics to weathering processes of soil minerals first appeared in the work of Wollast (1967). He concluded that the rate-limiting step for weathering of feldspars was diffusion (Chapter 7). This work touched off a lively debate that is still raging today about whether weathering of feldspars and ferromagnesian minerals is controlled by chemical reaction (CR) or diffusion.

Many of the early studies on kinetics of soil chemical processes were obviously concerned with diffusion-controlled exchange phenomena that had half-lives ($t_{1/2}$) of 1 s or greater. However, we know that time scales for soil chemical processes range from days to years for some weathering processes, to milliseconds for degradation, sorption, and desorption of certain pesticides and organic pollutants, and to microseconds for surface-catalyzed like reactions. Examples of the latter include metal sorption–desorption reactions on oxides.

To study rapid reactions, traditional batch and flow techniques are inadequate. However, the development of stopped flow, electric field pulse, and particularly pressure-jump relaxation techniques have made the study of rapid reactions possible (Chapter 4). German and Japanese workers have very successfully studied exchange and sorption–desorption reactions on oxides and zeolites using these techniques. In addition to being able to study rapid reaction rates, one can obtain chemical kinetics parameters. The use of these methods by soil and environmental scientists would provide much needed mechanistic information about sorption processes.

Environmental quality issues are receiving major attention in the soil and environmental sciences. To properly understand the fate of pesticides and organic and inorganic pollutants in soils and lakes, one must have a knowledge of the rates and mechanisms of these reactions. Some work has appeared on these topics (Chapter 6), but obviously there is a veritable need for more studies.

Perhaps one of the most unknown areas in kinetics of soil chemical processes is redox dynamics (Chapter 8). Some work on reductive dissolution of manganese oxides [Mn(III/IV)] and oxidation of As(V), Cr(III), and Pu(III/IV) by oxides has appeared. However, a comprehensive understanding of redox kinetics in heterogeneous systems is lacking.

Finally, the determination of rate parameters for soil chemical processes is fundamental if accurate and complete models are to be developed (Chapter 9) that will predict the fate of ions, pesticides, and organic pollutants in soil and aqueous systems.

In short, much future research on kinetics of soil chemical processes is needed. Areas worthy of investigation include improved methodologies, increased use of spectroscopic and rapid kinetic techniques to determine mechanisms of reactions on soils and soil constituents, kinetic modeling, kinetics of anion reactions, redox and weathering dynamics, kinetics of ternary exchange phenomena, and rates of organic pollutant reactions in soils and sediments.

2

Application of Chemical Kinetics to Soil Chemical Reactions

INTRODUCTION

The application of chemical kinetics to even homogeneous solutions is often arduous. When kinetic theories are applied to heterogeneous soil constituents, the problems and difficulties are magnified. With the latter in mind, one must give definitions immediately for two terms—kinetics and chemical kinetics. *Kinetics* is a general term referring to time-dependent phenomena. *Chemical kinetics* can be defined as the "study of the rate of chemical reactions and of the molecular processes by which reactions occur where transport is not limiting" (Gardiner, 1969). In soil systems, many kinetic processes are a combination of both chemical kinetics or reaction-controlled kinetics, and transport-controlled kinetics. In fact, many of the studies conducted thus far on time-dependent behavior of soils and soil constituents have been involved with transport-controlled kinetics and not chemical kinetics. The reasons for this are discussed later.

There are two salient reasons for studying the rates of soil chemical processes: (1) to predict how quickly reactions approach equilibrium or quasi-state equilibrium, and (2) to investigate reaction mechanisms. There are a number of excellent books on chemical kinetics (Laidler, 1965; Hammes, 1978; Eyring *et al.*, 1980; Moore and Pearson, 1981) and chemical engineering kinetics (Levenspiel, 1972; Froment and Bischoff, 1979) that the reader may want to refer to. The purpose of this chapter is to apply principles of chemical kinetics as discussed in the preceding books to soil chemical processes.

RATE LAWS

Differential Rate Laws

To fully understand the kinetics of soil chemical reactions, a knowledge of the rate equation or rate law explaining the reaction system is required.

By definition, a rate equation or law is a differential equation. In the following reaction (Bunnett, 1986),

$$a\mathrm{A} + b\mathrm{B} \longrightarrow y\mathrm{Y} + z\mathrm{Z} \tag{2.1}$$

the rate is proportional to some power of the concentrations of reactants A and B and/or other species (C, D, etc.) present in the system. The power to which a concentration is raised may equal zero (i.e., the rate may be independent of that concentration), even for reactant A or B.

For reactions occurring in liquid systems at constant volume, reaction rate is expressed as the number of reactant species (molecules or ions) changed into product species per unit of time and per unit of volume of the reaction system. Rates are expressed as a decrease in reactant concentration or an increase in product concentration per unit time. Therefore, if the substance chosen is reactant A, which has a concentration [A] at any time t, the rate is $(-d[\mathrm{A}])/(dt)$, while the rate with regard to a product Y having a concentration [Y] at time t is $(d[\mathrm{Y}])/(dt)$.

However, the stoichiometric coefficients in Eq. (2.1), a, b, y, and z, must also be considered. One can write

$$\frac{d[\mathrm{Y}]/dt}{y} = \frac{-d[\mathrm{A}]/dt}{a} = k[\mathrm{A}]^{\alpha}[\mathrm{B}]^{\beta}\cdots \tag{2.2}$$

where k is the rate constant and α is the order of the reaction with respect to reactant A and can be referred to as a partial order. Similarly, the partial order β is the order with respect to B. These orders are experimental quantities and are not necessarily integral. The sum of all the partial orders, α, β, ... is referred to as the overall order (n) and may be expressed as,

$$n = \alpha + \beta + \cdots \tag{2.3}$$

Once the values of α, β, etc. are determined, the rate law is defined. Reaction order is an experimental quantity and conveys only information about the manner in which rate depends on concentration. One should not use order to mean the same as "molecularity," which concerns the number of reactant particles (atoms, molecules, free radicals, or ions) entering into an elementary reaction. An elementary reaction is one in which no reaction intermediates have been detected, or need to be postulated to describe the chemical reaction on a molecular scale. Until other evidence is found, an elementary reaction is assumed to occur in a single step and to pass through a single transition state (Bunnett, 1986). The stoichiometric coefficients in the denominators of the differentials of Eq. (2.2) guarantee that the equation represents the rate of reaction regardless of whether rate of consumption of a reactant or of formation of a product is considered.

Rate laws are determined by experimentation and cannot be inferred only by examining the overall chemical reaction equation (Sparks, 1986). Rate laws serve three primary purposes: (1) they permit the prediction of the rate, given the composition of the mixture and the experimental value of the rate constant or coefficient; (2) they enable one to propose a mechanism for the reaction; and (3) they provide a means for classifying reactions into various orders.

Kinetic phenomena in soil or on soil constituents can be described by employing mechanistic rate laws, apparent rate laws, apparent rate laws including transport processes, or mechanistic rate laws including transport (Skopp, 1986).

Mechanistic Rate Laws

Definition and Verification. The use of mechanistic rate laws to study soil chemical reactions assumes that only *chemical kinetics* phenomena are being studied. Transport-controlled kinetics which involve physical aspects of soils are ignored. Thus, with mechanistic rate laws, mixing and/or flow rates do not influence the reaction rate (Skopp, 1986).

The objective of a mechanistic rate law is to ascertain the correct fundamental rate law. The reaction sequence for determination of mechanistic rate laws may represent several reaction paths and steps either purely in solution or on the soil surface of a well-stirred dilute soil suspension. All processes represent fundamental steps of a chemical rather than a physical nature (Skopp, 1986).

Given the following elementary reaction between species A, B, and Y, the chemical equation is

$$\mathrm{A} + 2\mathrm{B} \underset{k_{-1}}{\overset{k_1}{\rightleftharpoons}} \mathrm{Y} \tag{2.4}$$

A forward reaction rate law can be written as

$$d[\mathrm{A}]/dt = -k_1[\mathrm{A}][\mathrm{B}]^2 \tag{2.5}$$

where k_1 is the forward rate constant.

The reverse reaction rate law for Eq. (2.4) can be expressed as

$$d[\mathrm{A}]/dt = +k_{-1}\,[\mathrm{Y}] \tag{2.6}$$

where k_{-1} is the reverse rate constant.

For *chemical kinetics* to be operational and thus Eqs. (2.5) and (2.6) to be valid, Eq. (2.4) must be an elementary reaction. To definitively determine this, one must prove experimentally that Eq. (2.4) and the rate law are valid.

To verify that Eq. (2.4) is indeed elementary, one can employ experimental conditions that are dissimilar from those used to ascertain the rate law. For example, if the *k* values change with flow rate, one is determining nonmechanistic or apparent rate coefficents. This was the case in a study by Sparks *et al.* (1980b), who studied the rate of potassium desorption from soils using a continuous flow method (Chapter 3). They found the apparent desorption rate coefficients (k'_d) increased in magnitude with flow rate (Table 2.1). Apparent rate laws are still useful to the experimentalist and can provide useful time-dependent information.

The determination of mechanistic rate laws for soil chemical processes is very difficult since microscopic heterogeneity is pronounced in soils and even for most soil constituents such as clay minerals, humic substances, and oxides. Heterogeneity can be enhanced due to different particle sizes, types of surface sites, etc. As will be discussed more completely in Chapter 3, the determination of mechanistic rate laws is also complicated by the type of kinetic methodology one uses. With some methods used by soil and environmental scientists, transport-controlled reactions are occurring and thus mechanistic rate laws cannot be determined.

TABLE 2.1 Effect of Flow Velocity on the Magnitude of the k'_d of the Ap and B22t Soil Horizons from Nottoway County[a]

Horizon	Flow velocity (ml min^{-1})	k'_d (h^{-1})	
		Al-saturated	Ca-saturated
Ap	0.0	0.83	1.11
	0.5	0.85	1.18
	1.0	0.87	1.23
	1.5	0.91	1.32
B22t	0.0	0.33	0.26
	0.5	0.37	0.28
	1.0	0.41	0.30
	1.5	0.48	0.34

[a] These k'_d values were obtained by plotting a regression line of the triplicate k'_d values (determined in triplicate experiments) versus flow velocity. The r values were 0.970 and 0.973 for the Ap and B22t horizons, respectively, which were significant at the 1% level of probability. From Sparks *et al.* (1980b), with permission.

Skopp (1986) has noted that Eq. (2.5) or (2.6) alone, are only applicable far from equilibrium. For example, if one is studying adsorption reactions near equilibrium, back or reverse reactions are occurring as well. The complete expression for the time dependence must combine Eqs. (2.5) and (2.6) such that,

$$d[A]/dt = -k_1[\mathrm{A}][\mathrm{B}]^2 + k_{-1}[\mathrm{Y}] \tag{2.7}$$

Equation (2.7) applies the principle that the net reaction rate is the difference between the sum of all reverse reaction rates and the sum of all forward reaction rates.

Determination of Mechanistic Rate Laws and Rate Constants. One can determine mechanistic rate laws and rate constants by analyzing data in several ways (Bunnett, 1986; Skopp, 1986). These include ascertaining initial rates, using integrated rate equations such as Eqs. (2.5)–(2.7) directly and graphing the data, and employing nonlinear least-square techniques to determine rate constants.

Graphical Assessment Using Integrated Equations Directly. Another way to ascertain mechanistic rate laws is to use an integrated form of Eq. (2.7). One way to solve Eq. (2.7) is to conduct a laboratory study and assume that one species is in excess (i.e., B) and therefore, constant. Mass balance relations are also useful. For example $[\mathrm{A}] + [\mathrm{Y}] = A_0 + Y_0$ where Y_0 is the initial concentration of product. One must also specify an initial condition to solve rate equations. For example, Eq. (2.7) can be solved by assuming that [B] is constant and $Y_0 = 0$ and letting the initial condition be specified by $A = A_0$ at $t = 0$. Dropping the brackets from Eq. (2.7) for the sake of simplicity one obtains (Skopp, 1986):

$$A/A_0 = \{k_{-1} + k_c \exp[-t(k_c + k_{-1})]\}/(k_c + k_{-1}) \tag{2.8}$$

where $k_c = k_1 B^2$.

Equation (2.5) can also be integrated using the same initial conditions and one obtains a first-order equation (Skopp, 1986):

$$A/A_0 = \exp(-k_c t) \tag{2.9}$$

If Eq. (2.9) is appropriate, a graph of log (A/A_0) vs. t should yield a straight line with a slope equal to $-k_c$. However, based on this result alone, it is tenuous to conclude that Eq. (2.5) is the *only* possible interpretation of the data and that a straight-line graph indicates a first-order reaction. One can make these conclusions only if *no* other reaction mechanisms result in such a graphical relationship.

Similar arguments hold for the integrated form of Eq. (2.6) when $Y = Y_0$ at $t = 0$ and $A = Y_0 - Y$ such that

$$A/Y_0 = 1 - \exp(k_{-1}t) \tag{2.10}$$

The solution to Eq. (2.7) assuming [B] is constant and $A_0 = 0$ is (Skopp, 1986),

$$A/Y_0 = [k_{-1}/(k_c + k_{-1})]\{1 - \exp[-t(k_c + k_{-1})]\} \tag{2.11}$$

Graphs of log $(1 - A/Y_0)$ vs. t are commonly used to test the validity of Eq. (2.10). However, Eq. (2.11), like Eq. (2.8), shows more complex behavior than simple graphical methods reveal. Thus, one should be cautious about making definitive statements concerning rate constants and particularly mechanisms, based solely on data according to integrated equations like those in Eqs. (2.9) and (2.10) unless other reaction mechanisms have been ruled out.

Often when time-dependent data are plotted using an equation for a particular reaction order, curvature results. There are several explanations for this. It can be caused by an incorrect assumption of reaction order. For example, if first-order kinetics is assumed but the reaction is second-order, downward curvature is observed (Bunnett, 1986). If second-order kinetics is assumed but the reaction is really first-order, upward curvature results. Curvature could also be due to fractional, third, or higher reaction orders or to mixed reaction orders.

If a reaction progresses to a state of equilibrium that is short of completion, a kinetic plot based on the assumption that the reaction went to completion shows downward curvature with an eventual zero slope. Curvature can also be caused by temperature changes during an experiment. Decreasing temperature causes downward curvature, and increasing temperature results in upward curvature (Bunnett, 1986). These changes can result if temperature is not held constant during an experiment, or if the temperature of the sorptive solution is not the same as that used in the kinetic run.

Curvature can also be caused by side reactions. However, these reactions do not always cause deviations from linearity (Bunnett, 1986). This is another reason one should not make definitive conclusions about a linear kinetic plot.

The graphical method of determining k values from integrated equations works well if the points closely approximate a straight line or if they scatter randomly. Sometimes one can draw a straight line through every point; thus, the slope of the line is adequate for evaluation of k (Bunnett, 1986).

Initial Rate Method. Using integrated equations like Eqs. (2.5), (2.6), or (2.7) to directly determine a rate law and rate constants is risky. This is particularly true if secondary or reverse reactions are important in equations like (2.5) and (2.6). One sound option is to establish these equations directly using initial rates (Skopp, 1986).

With this method, the concentration of a reactant or product is plotted versus time for a very short initial period of the reaction during which the concentrations of the reactants change so little that the instantaneous rate is hardly affected (Bunnett, 1986).

The initial rate is the limit of the reaction rate as time reaches zero. Using the initial rate method, one could ascertain Eqs. (2.5) and (2.6) by finding out how the initial rates ($\lim d[\mathrm{A}]/dt$) depend on the initial concentrations (A, B, Y). Experiments are conducted such that initial concentrations of each reactant are altered while the other concentrations are constant. It is desirable with this method to have one reactant in much higher concentration than the other reactant(s).

With the initial rate method, one must use an extremely sensitive analytical method to determine product concentrations (Bunnett, 1986). Titration methods may not be suitable, particularly if low levels of product concentration are present. Therefore, physical techniques such as spectrophotometry or conductivity are utilized.

Least-Squares Techniques. The value of k can also be obtained using least-squares techniques. This statistical method fits the best straight line to a set of points that are supposed to be linearly related. The formula for a straight line is

$$y = mx + b \qquad (2.12)$$

The most tractable form of least-squares analysis assumes that values of the independent variable x are known without error and that experimental error is manifested only in values of the dependent variable y. Most kinetic data approximate this situation, since the times of observation are more accurately measurable than the chemical or physical quantities related to reactant concentrations (Bunnett, 1986). The straight line selected by least-squares analysis is that which minimizes the sum of the squares of the deviations of the y variable from the line.

The slope m and intercept b can be calculated by least-squares analysis using Eqs. (2.13) and (2.14), respectively.

$$\text{Slope} = m = \frac{n\Sigma xy - \Sigma x \Sigma y}{n\Sigma x^2 - (\Sigma x)^2} \tag{2.13}$$

$$\text{Intercept} = b = \frac{\Sigma y \Sigma x^2 - \Sigma x \Sigma xy}{n\Sigma x^2 - (\Sigma x)^2} \tag{2.14}$$

where n is the number of data points and the summations are for all data points in the sets. For further information on least-squares analysis, one can consult any number of textbooks including those of Draper and Smith (1981) and Montgomery and Peck (1982).

Apparent Rate Laws

Apparent rate laws include both chemical kinetics and transport-controlled processes. One can ascertain rate laws and rate constants using the previous techniques. However, one does not need to prove that *only* elementary reactions are being studied (Skopp, 1986). Apparent rate laws indicate that diffusion or other microscopic transport phenomena affect the rate law (Fokin and Chistova, 1967). Soil structure, stirring, mixing, and flow rate all affect the kinetic behavior when apparent rate laws are operational.

Transport with Apparent Rate Law

A fourth type of rate law, transport with apparent rate law, is a form of apparent rate law that includes transport processes. This type of rate-law determination is ubiquitous in the modeling literature (Cho, 1971; Rao *et al.*, 1976; Selim *et al.*, 1976a; Lin *et al.*, 1983). Kinetic-based transport models are more fully described in Chapter 9. With these rate laws, transport-controlled kinetics are emphasized more and chemical kinetics less. The apparent rate can often depend on water flux (Skopp and Warrick, 1974; Overman *et al.*, 1980) or other physical processes. One also usually assumes either first- or zero-order kinetics is operational.

Transport with Mechanistic Rate Laws

Here one makes an effort to describe simultaneously transport-controlled and chemical kinetics processes (Skopp, 1986). Thus, an attempt is made to describe both the chemistry and physics accurately. For example, outflow curves from miscible displacement experiments on soil columns are matched to solutions of the conservation of mass equation. The matching process introduces a potential ambiquity such that experimental uncertainties are translated into model uncertainties. Often, an error in the description of the physical process is compensated for by an error in the chemical process and vice-versa (i.e., Nkedi-Kizza *et al.*, 1984).

EQUATIONS TO DESCRIBE KINETICS OF REACTIONS ON SOIL CONSTITUENTS

Introduction

A number of equations have been used to describe the kinetics of soil chemical processes (see, e.g., Sparks, 1985, 1986). Many of these equations offer a means of calculating rate coefficients, which then can be used to determine energies of activation (E), which reveal information concerning rate-limiting steps. Energies of activation measure the magnitude of forces that must be overcome during a reaction process, and they vary inversely with reaction rate.

However, as noted earlier, conformity of kinetic data to a particular equation does not necessarily mean it is the best model, nor can one propose mechanisms based on this alone.

First-Order Reactions

Derivations. According to the usual convention, one lets *a* represent the initial concentration $[A]_0$, of species A, *b* the initial concentration $[B]_0$, of species B, and *y* the concentration of product Y or Z [see Eq. (2.1)] at any time (Bunnett, 1986). For a first-order reaction,

$$\frac{dy}{dt} = k[\text{A}] = k(a - y) \tag{2.15}$$

Rearranging,

$$\frac{dy}{a - y} = k\,dt \tag{2.16}$$

Integrating,

$$-\ln(a - y) + \ln a = kt \tag{2.17}$$

Or, using base 10 logarithms,

$$-\log(a - y) + \log a = \frac{k}{2.30}t \tag{2.18}$$

Eq. (2.17) can also be written as,

$$-\ln[\text{A}] + \ln[\text{A}]_0 = kt \tag{2.19}$$

Equation (2.17), (2.18), or (2.19) indicates that a plot of the negative of the logarithm of [A] or of $(a - y)$ versus time should be a straight line with slope k or $k/2.30$. As noted earlier (Section IIB, 2b), obtaining such a linear plot from experimental data is a necessary but not sufficient condition for one to conclude that the reaction is kinetically first-order. Even if the kinetic plot using a first-order equation is linear over 90% of the reaction, deviations from the assumed rate expression may be hidden (Bunnett, 1986). When other tests confirm that it is first-order, the rate constant k, is either the negative of the slope [Eq. (2.17) or (2.19)] or 2.30 times the negative of the slope [Eq. (2.18)].

One way to test for first-order behavior is to carry out the rate determination at another initial concentration of reactant A, such as double or half the original, but preferably 10-fold or smaller (Bunnett, 1986). If the reaction is first-order, the slope according to Eq. (2.17) or (2.18) should be unchanged. It is also necessary to show that reaction rate is not affected by a species whose concentrations do not change considerably during a reaction run; these may be substances not consumed in the reaction (i.e., catalysts) or present in large excess (Bunnett, 1986).

The half-life ($t_{1/2}$) of a reaction is the time required for half of the original reactant to be consumed. A first-order reaction has a half-life that is related only to k and is independent of the concentration of the reacting species. After one half-life, $(a - y)$ equals $a/2$, and Eq. (2.17) can be rewritten as,

$$-\ln \frac{a}{2} + \ln a = kt_{1/2}$$

Consolidating and rearranging,

$$t_{1/2} = \frac{\ln 2}{k} = \frac{0.693}{k} \tag{2.20}$$

The half-life depends on reactant concentration and becomes longer the less concentrated the reactant. Thus, it can take a long time to reach a satisfactory infinity value for a second-order reaction.

Application of First-Order Reactions to Soil Constituents. Many investigations on soil chemical processes have shown that first-order kinetics describe the reaction(s) well. Single or multiple first-order reactions have been observed for ionic reactions involving: As(III) (Oscarson *et al.*, 1983), potassium (Mortland and Ellis, 1959; Burns and Barber, 1961; Reed and Scott, 1962; Huang *et al.*, 1968; Sivasubramaniam and Talibudeen, 1972; Jardine and Sparks, 1984; Ogwada and Sparks, 1986b), nitrogen (Stanford *et al.*, 1975; Kohl *et al.*, 1976; Carski and Sparks, 1987), phosphorus (Amer *et al.*, 1955; Griffin and Jurinak, 1974; Li *et al.*, 1972; Vig *et al.*, 1979), copper (Jopony and Young, 1987), lead (Salim and Cooksey, 1980), cesium (Sawhney, 1966), boron (Griffin and Burau, 1974; Carski and Sparks, 1985), sulfur (Hodges and Johnson, 1987), aluminum (Jardine and Zelazny, 1986), and chlorine (Thomas, 1963). First-order equations have also been used to describe molecular reactions on soils and soil constituents, including pesticide interactions (Walker, 1976a,b; Rao and Davidson, 1982; McCall and Agin, 1985). Data from these studies have been fitted to first-order equations by methods described earlier.

Sparks and Jardine (1984) studied the kinetics of potassium adsorption on kaolinite, montmorillonite, and vermiculite (Fig. 2.1) and found that a single first-order reaction described the data well for kaolinite and smectite while two first-order reactions described adsorption on vermiculite. One will note deviations from first-order kinetics at longer time periods, particularly for montmorillonite and vermiculite, because a quasi-equilibrium state is reached. These deviations result because first-order equations are only applicable far from equilibrium (Skopp, 1986); back reactions could be occurring at longer reaction times.

Griffin and Jurinak (1974) studied B desorption kinetics from soil and observed two separate first-order reactions and one very slow reaction. They postulated that the two first-order reactions were due to desorption from two independent B retention sites associated with hydroxy-Al, −Fe,

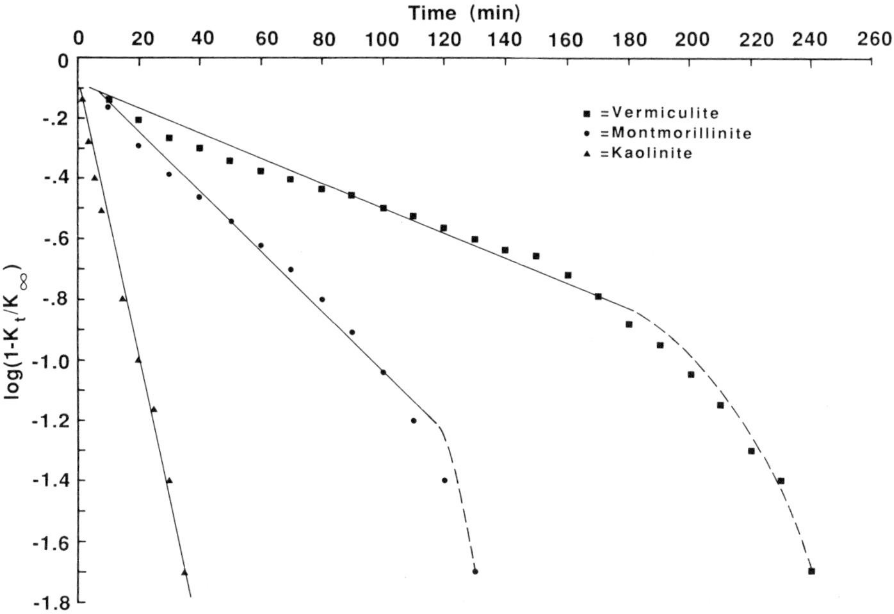

Figure 2.1. First-order plots of potassium adsorption on clay minerals where K_t is quantity of potassium adsorbed at time t and K_∞ is quantity of potassium adsorbed at equilibrium. [From Sparks and Jardine (1984), with permission.]

and −Mg materials in the clay fraction of the soils. The third or lowest reaction rate was attributed to diffusion of B from the interior of clay minerals to the solution phase.

There are dangers, however, in attributing multiple slopes, obtained from plotting time-dependent data according to various kinetic equations, to different sites for reactivity. This is particularly true when the only evidence for such conclusions is multiple slopes. Even if one finds, for example, that data are best described by two first-order reactions, one should not then conclude that two mechanisms are operable. Such conclusions are analogous to deciding that multiple slopes obtained with the Langmuir equation are indicative of different sorption sites and mechanisms (see, e.g., Harter and Smith, 1981) One should refrain from making such judgments unless other lines of evidence also point to multiple reaction sites.

There are several ways to determine kinetically that multiple first-order or other reaction order slopes are present, and that they indicate different sites or mechansims for sorption. One could determine rate-limiting steps (Chapter 5), E values could be measured, or materials that affect specific

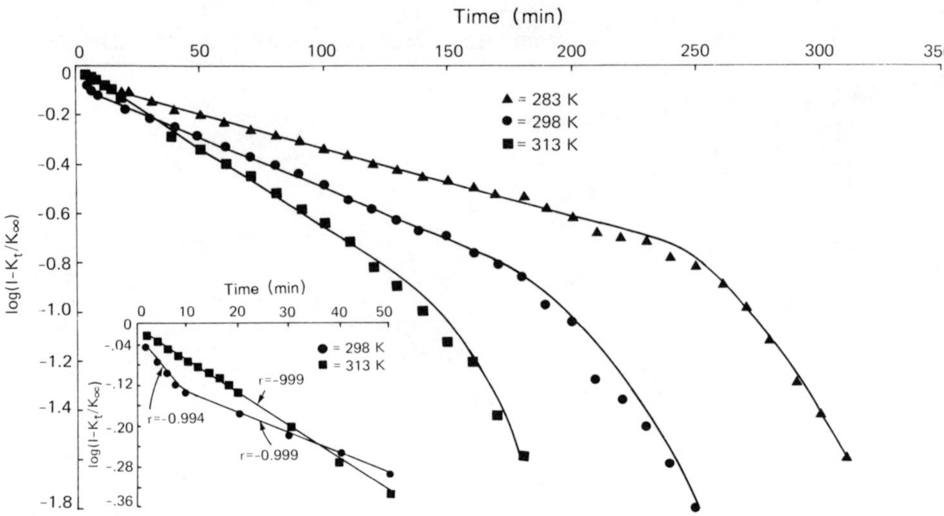

Figure. 2.2. First-order kinetics for potassium adsorption at three temperatures on Evesboro soil, with inset showing the initial 50 min of the first-order plots at 298 and 313 K. Terms are defined in Fig. 1. [From Sparks and Jardine (1984), with permission.]

colloidal sites (blocking agents) could be used to isolate different sorption sites.

An example of the last one can be found in the work of Jardine and Sparks (1984). They found that potassium adsorption and desorption in a soil conformed well to first-order reactions at 283 and 298 K and that two apparent simultaneous first-order reactions existed (Fig. 2.2). The first slope contained both a rapid reaction (rxn 1) and a slow reaction (rxn 2). The second slope described only rxn 2. The difference between the two slopes yielded the slope for rxn 1. The first reaction conformed to first-order kinetics for about 8 min, after which time a second apparent reaction proceeded for many hours. Sparks and Jardine (1984) used different blocking agents, including cetyltrimethylammonium bromide (CTAB), which sorbed only on external surface sites (Fig. 2.3), to show that the two slopes were describing two reactions on different sites for potassium adsorption–desorption. Based on the CTAB results, rxn 1 was ascribed to external surface sites of the organic and inorganic phases of the soil that were readily accessible for cation exchange. Reaction 2 was attributable to less accessible sites of organic matter and interlayer sites of the 2 : 1 clay minerals such as vermiculitic clays that predominated in the < 2 μm clay fraction.

Another way to more directly prove or disprove mechanisms based on different time-dependent slopes is to use spectroscopic techniques.

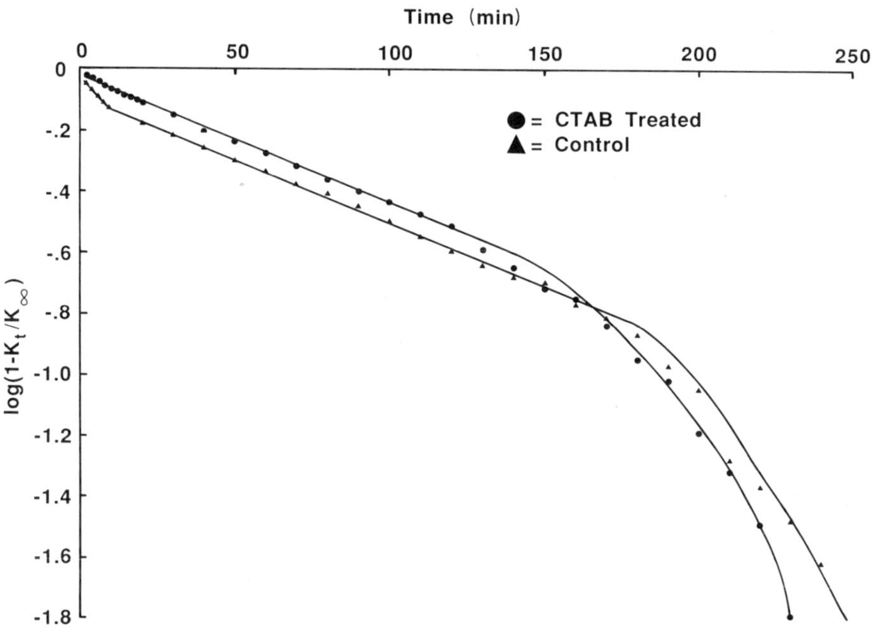

Figure 2.3. First-order kinetics for potassium adsorption at 298 K on Evesboro soil treated with cetyltrimethylammonium bromide (CTAB). Terms are defined in Fig. 1. [From Jardine and Sparks (1984), with permission.]

Methods such as nuclear magnetic resonance (NMR), electron spectroscopy for chemical analysis (ESCA), electron spin resonance (ESR), infrared (IR), and laser raman spectroscopy could be used in conjunction with rate studies to define mechanisms. Another alternative would be to use fast kinetic techniques such as pressure-jump relaxation, electric field pulse, or stopped flow (Chapter 4), where chemical kinetics are measured and mechanisms can be definitively established.

Other Reaction-Order Equations

Zero-Order Reactions. Zero-order reactions have been applied to describe potassium (Mortland, 1958; Burns and Barber, 1961), chromium (Amacher and Baker, 1982), and nitrogen reactions in soils (Patrick, 1961; Broadbent and Clark, 1965; Keeney, 1973).

Second-Order Reactions. If one considers a reaction according to Eq. (2.1), which is overall second-order but first-order in A and first-order in B, then according to the symbolism used earlier (Bunnett, 1986),

$$\frac{dy}{dt} = k(a - y)(b - y) \tag{2.21}$$

Rearranging,

$$\frac{dy}{(a - y)(b - y)} = k\,dt$$

Integrating,

$$\frac{1}{a - b} \ln \frac{b(a - y)}{a(b - y)} = kt \tag{2.22}$$

Equation (2.22) is valid only if $a \neq b$. An alternate form of Eq. (2.22) is

$$\ln \frac{a - y}{b - y} + \ln \frac{b}{a} = (a - b)kt \tag{2.23}$$

or,

$$\log \frac{a - y}{b - y} + \log \frac{b}{a} = \frac{(a - b)kt}{2.30} \tag{2.24}$$

Thus, plots of the logarithm of $[(a - y)/(b - y)]$ versus time should be linear with slopes $(a - b)k$ or $(a - b)k/2.30$, depending on which type of logarithm is used.

If the experiment is arranged so that the initial concentrations of A and B are equal or if the reaction is second-order in reactant A, Eq. (2.21) becomes,

$$\frac{dy}{dt} = k(a - y)^2 \tag{2.25}$$

Upon rearrangement and integration,

$$\frac{1}{a - y} - \frac{1}{a} = kt \tag{2.26}$$

a plot of the reciprocal of $(a - y)$ versus time is linear with slope k.

Whereas the half-life for a first-order reaction is independent of reactant concentration, that for a second-order reaction is not. If one inserts $a/2$ for $(a - y)$ in Eq. (2.26), one obtains

$$t_{1/2} = \frac{1}{ak} \tag{2.27}$$

Phosphate reactions on calcite (Kuo and Lotse, 1972; Griffin and Jurinak, 1974) have been described using second-order reactions. Also, recent work on Al reactions in soils has employed second-order reactions (Jardine and Zelazny, 1986).

Kuo and Lotse (1972) derived a second-order equation, which is presented below, and used this equation to describe the rate of PO_4 sorption on $CaCO_3$ and Ca-kaolinite. This equation considered both the change in PO_4 concentration in solution and the surface saturation of the sorbent during the sorption process. This equation can be writtten as

$$\frac{dq}{dt} = k_1(C_0 - q)(M - q) - k_{-1}q \tag{2.28}$$

where q is the quantity of ions sorbed, $(C_0 - q)$ is the concentration of ions remaining in solution where C_0 is the initial concentration of ions, and $(M - q)$ is the surface unsaturation.

At equilibrium (eq), dq/dt will equal zero. Then,

$$k_1(C_0 - q)_{eq}(M - q)_{eq} = k_{-1}q_{eq} \tag{2.29}$$

By arranging Eq. (2.29), and expressing q and M in mol kg^{-1} of sorbent rather than in mol ion l^{-1}, the Langmuir equation is obtained,

$$\frac{1}{K_{eq}M} = \frac{C_{eq}}{q} - \frac{C_{eq}}{M} \tag{2.30}$$

where K_{eq} is the equilibrium constant and C_{eq} is the equilibrium concentration of ions. By integrating Eq. (2.28), one obtains

$$\ln\left(\frac{q - A - B}{q + B - A}\right) = 2Ak_1t + \ln\left(\frac{B + A}{B - A}\right) \tag{2.31}$$

where

$$A = \left[\frac{1}{4}\left(C_0 + M + \frac{k_{-1}}{k_1}\right)^2 - C_0M\right]^{1/2} \tag{2.32}$$

and

$$B = \frac{1}{2}\left(C_0 + M + \frac{k_{-1}}{k_1}\right) \tag{2.33}$$

The parameters A and B are constants and contain a concentration unit. By plotting $\ln(q - A - B/q + A - B)$ as a function of t, a straight line is obtained with a slope equal to $2Ak_1$. Kuo and Lotse (1972) found that the second-order rate constant decreased with increasing phosphorus concentration in the $CaCO_3$–PO_4 system, which they explained using the Brønsted–Bjerrum activity-rate theory of ionic reactions in dilute solutions. The Brønsted equation states that "the logarithm of the rate coefficient is inversely proportional to the square root of the ionic strength, when the reaction between the two molecules involves charges of different sign" (Kuo and Lotse, 1972). Kuo and Lotse (1972) determined that k_1 for PO_4 sorption on Ca-kaolinite increased with increasing PO_4 concentration. Novak and Adriano (1975) found that a second-order equation like that given by Kuo and Lotse (1972) described phosphorus kinetics better than other models.

Reactions of Higher or Fractional Order. For reactions of order α in reactant A and of zero-order in other species,

$$\frac{-d[A]}{dt} = k[A]^{\alpha} \tag{2.34}$$

Letting $[A]_0$ represent the initial concentration of A and $[A]$ its concentration at any time, one may integrate (except when $\alpha = 1$) to obtain, (Bunnett, 1986).

$$\frac{1}{n-1}\left(\frac{1}{[A]^{\alpha-1}} - \frac{1}{[A]_0^{\alpha-1}}\right) = kt \tag{2.35}$$

Eq. (2.26) is the special case of Eq. (2.35) for $n = 2$.

When the order is $\frac{1}{2}$, $\frac{3}{2}$, and 3, Eq. (2.35) assumes the form of Eqs. (2.36), (2.37) and (2.38), respectively.

$$\sqrt{[A]_0} - \sqrt{[A]} = \frac{kt}{2} \quad \text{(half-order)} \tag{2.36}$$

$$\frac{1}{\sqrt{[A]}} - \frac{1}{\sqrt{[A]_0}} = \frac{kt}{2} \quad \text{(three-halves order)} \tag{2.37}$$

$$\frac{1}{[A]^2} - \frac{1}{[A]_0^2} = 2\,kt \quad \text{(third-order)} \tag{2.38}$$

Equations (2.35), (2.37), and (2.38) are also obtained if the reaction is of order $\alpha - 1$ in reactant A and of order one in B, and if the initial concentrations of A and B (and maybe other reactants) are in the ratio of their stoichiometric coefficients.

Often fractional orders best describe soil chemical processes. For example, the reaction order for dissolution of oxides, calcite, feldspars, and ferromagnesian minerals is often <1 (Stumm *et al.*, 1985; Bloom and Erich, 1987).

Two-Constant Rate Equation

Kuo and Lotse (1973) used a two-constant rate equation, derived below, which is adapted from the Freundlich equation to study the kinetics of PO_4 sorption and desorption on hematite and gibbsite.

A kinetic equation was developed by inserting a time-dependent expression into the Freundlich equation. The Freundlich equation can be written as,

$$q = k_F C_f^{c/d} \tag{2.39}$$

where k_F is the Freundlich constant, C_f is the final sorbate concentration, and c and d are both integers with $d > c$.

Kuo and Lotse (1973) presumed that the slope of a Freundlich plot was time independent. The physical meaning of the concentration term exponent in the Freundlich equation is unclear, but has generally been <1 and related to the characteristics of the sorbent. However, the exponent is time-independent, whereas the intercept is time-dependent.

The expression $(1 - e^{-k_2 t})$ can be inserted into Eq. (2.39) such that

$$q^d = k_1(1 - e^{-k_2 t})C_f \tag{2.40}$$

where k_1 and k_2 are constants, $k_1[1 - e^{-k_2 t}] = k^d$

For $t = 0$, $q = 0$, and $t \to \infty$, Eq. (2.40) is reduced to Eq. (2.39). From Eq. (2.40) one sees that changes in t can only affect the intercept in the Freundlich plot. Rearrangement and successive transformations give,

$$\frac{q}{C_f^{c/d}} = k_1^{1/d}(1 - e^{-k_2 t})^{1/d} \tag{2.41}$$

and

$$\begin{aligned} \ln\left(\frac{q}{C_f^{c/d}}\right) &= \ln k_1^{1/d} + 1/d \ln(1 - e^{-k_2 t}), \\ &= 1/d \ln k_1 + \frac{1}{d} \ln[(1 - (1 - k_2 t) + \eta] \end{aligned} \tag{2.42}$$

in which η is an error term and equals,

$$\eta = \frac{1}{2!}(k_2 t)^2 - \frac{1}{3!}(k_2 t)^3 + \cdots$$

Equation (2.42) contains a total of five unknown constants. To solve the equation, a trial value of the constant k_2 is needed. However, if the constant k_2 (units of h^{-1}) is relatively small, the net effect of η in Eq. (2.42) can be neglected. Accordingly, Eq. (2.42) can be simplified,

$$\ln\left(\frac{q}{C_f^{c/d}}\right) = 1/d \ln k_1k_2 + 1/d \ln t \tag{2.43}$$

$$q = A^{1/d}C^{c/d}t^{1/d} \tag{2.44}$$

where $A = k_1k_2$. Since the values of the (c/d) term are $\ll 1$, the change of $C_f^{c/d}$ as a function of time is small. Then $C_f^{c/d}$ can be considered a constant, and Eq. (2.44) is rewritten as

$$q = KC_o t^{1/d} \tag{2.45}$$

where K is a constant

Kuo and Lotse (1973) plotted PO_4 sorbed versus t on a log–log scale and calculated $(1/d)$ and K from the slope and intercept, respectively, of the straight line.

The two-constant rate equation was also used to describe PO_4 desorption from soil (Dalal, 1974), *K-Ca* exchange on soils (Sparks *et al.*, 1980a), and recently, by Jopony and Young (1987) to study the kinetics of copper desorption from soil and clay minerals.

Elovich Equation

The Elovich equation is one of the most widely used equations to describe the kinetics of heterogeneous chemisorption of gases on solid surfaces (Low, 1960). This equation assumes a heterogeneous distribution of adsorption energies where the E increases linearly with surface coverage (Low, 1960). Parravano and Boudart (1955) criticized using the Elovich equation for describing one unique mechanism since they found that it described a number of different processes, such as bulk or surface diffusion and activation and deactivation of catalytic surfaces. Recent theoretical studies on adsorption–desorption phenomena in oxide–aqueous solution systems illustrated that the applicability and method of fitting kinetic data to the Elovich equation requires accurate data at short reaction times (Aharoni and Ungarish, 1976, 1977). Ungarish and Aharoni (1981) have also pointed out the inappropriateness of the Elovich equation at very low and very high surface coverages (Atkinson *et al.*, 1970; Sharpley, 1983). These types of situations could well exist in soils or on soil constituent systems.

The Elovich equation has been used to describe the kinetics of PO_4 sorption and desorption on soils and soil minerals (Atkinson *et al.*, 1970; Chien and Clayton, 1980; Chien *et al.*, 1980; Sharpley, 1983), potassium reactions in soils (Sparks *et al.*, 1980b; Martin and Sparks, 1983; Sparks and Jardine, 1984; Havlin and Westfall, 1985), borate dissolution from soils (Peryea *et al.*, 1985), sulfur sorption and desorption kinetics in soils (Hodges and Johnson, 1987), and arsenite sorption on soils (Elkhatib) *et al.*, 1984a). In several of these studies, the Elovich equation was claimed to be superior to other kinetic equations based on high r and low SE (standard error) values (Chien and Clayton, 1980; Hodges and Johnson, 1987).

A form of the Elovich equation as applied to the adsorption of gases onto solid surfaces is

$$dq/dt = Xe^{-Yq} \tag{2.46}$$

where q is the amount sorbed at time t and X and Y are constants in a given experiment. The integrated form of Eq. (2.46) is

$$q = (1/Y)\ln(XY) + (1/Y)\ln(t + t_0) \tag{2.47}$$

where t_0 is the integration constant. One can make either of two assumptions when studying t_0 (Polyzopoulous *et al.*, 1986) The first assumption is that $t_0 \simeq 0$, which indicates that no other processes besides Elovichian ones are occurring; that is, the boundary condition $q = 0$ at $t = 0$ applies. Thus, with the assumption that $t_0 \simeq 0$, then

$$q = (1/Y)\ln(XY) + (1/Y)\ln(t) \tag{2.48}$$

Equation (2.48) is a simplified Elovich equation that several investigators have used to study the rates of soil chemical processes. An application of this equation to PO_4 sorption on soils is shown in Fig. 2.4, and one sees a

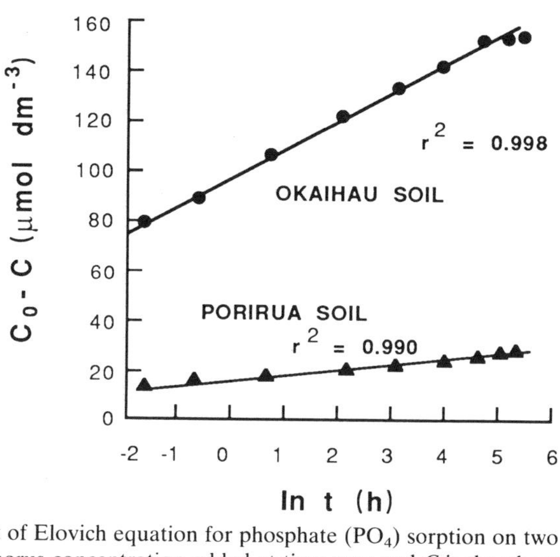

Figure 2.4. Plot of Elovich equation for phosphate (PO_4) sorption on two soils where C_0 is the initial phosphorus concentration added at time zero and C is the phosphorus concentration in the soil solution at time t. [From Chien and Clayton (1980), with permission.]

linear relationship. Chien and Clayton (1980) obtained the simplified equation given in Eq. (2.41) by using the relationship

$$q = (1/Y)\ln(1 + XYt) \tag{2.49}$$

from Eq. (2.47) and by assuming that the boundary condition $q = 0$ at $t = 0$ applies, and then by making the assumption that $XYt \gg 1$. The latter assumption follows directly from the original assumption $q = 0$ at $t = 0$, which when used with Eq. (2.47) leads to $t_0 = 1/XY$, that is, $XY \to \infty$ when $t_0 = 0$ (Polyzopoulous *et al.*, 1986). The second assumption is that $t_0 \neq 0$. Therefore, two possibilities exist.

First, the boundary condition $q = q_0$ at $t = 0$ is applicable. One can then test t_0 by determining its value graphically, numerically, or by regression so that it creates a linear relationship when q versus $\ln(t + t_0)$ is plotted. Thus, one is assuming an instantaneous pre-Elovichian process (Aharoni and Ungarish, 1976). However, studies have shown that a plot of q versus $\ln(t + t_0)$ is usually concave toward the q axis. Moreover, to linearize such a plot indicates that experimental data are Elovichian when in fact they may not be.

Aharoni and Ungarish (1976) suggested that another possibility was to introduce the boundary condition that $q = q_c$ at $t = t_c$ (q_c and $t_c > 0$) and thus the pre-Elovichian rate was finite (Fig. 2.5). They also give a procedure for estimating t_0 from experimental data without assuming anything *a priori* about a pre-Elovichian process (Polyzopoulous *et al.*, 1986). One can differentiate and rearrange Eq. (2.47) to yield a relationship

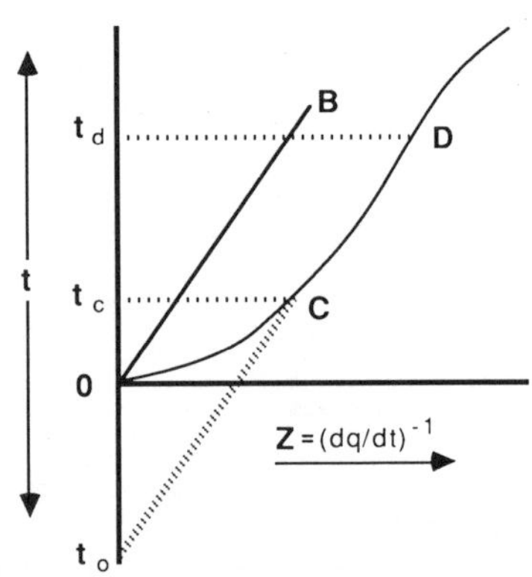

Figure 2.5. Schematic representation of t versus Z plots, where $Z = (dq/dt)^{-1}$ (adapted from Aharoni and Ungarish, 1976). The OCD general shape of the plots is obtained experimentally: 0C is the pre-Elovichian section; CD is the Elovichian section between $t = t_c$ and $t = t_d$, whose t intercept gives t_0; beyond D is the post-Elovichian section; and 0B is the plot when $t_0 = 0$. [From Polyzopoulous *et al.* (1986), with permission.]

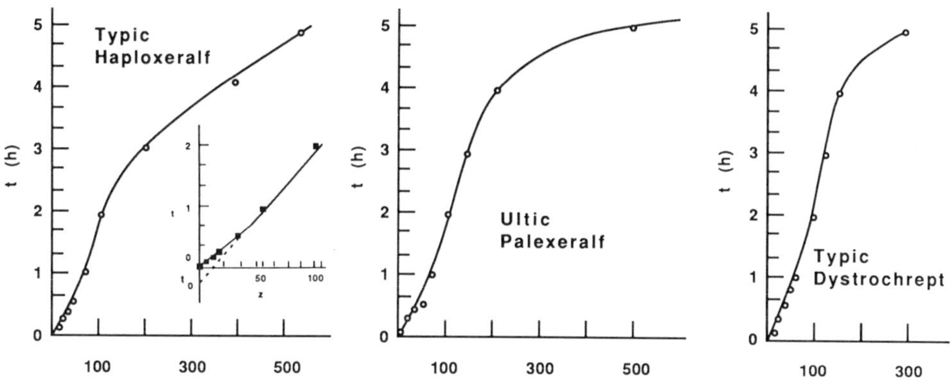

Figure 2.6. Plot of t versus Z for phosphate (PO_4) sorption on three Greek soils. [From Polyzopoulous *et al.* (1986), with permission.]

between adsorption and time t that is given in Eq. (2.50),

$$t = Z/Y - t_0 \tag{2.50}$$

where Z is $(dq/dt)^{-1}$ or the reciprocal of the rate. If Eq. (2.50) is valid, t is proportional to Z in any range and extrapolating to $Z = 0$ gives t_0 as an imaginary negative time (Polyzopoulous *et al.*, 1986). At this time the rate is infinity, and as it decreases with increasing q using Eq. (2.46), it coincides with the actual Elovichian rate (Fig. 2.5). Here, t values are given on the ordinate and the y intercept yields t_0 directly.

Equation (2.50) was used by Polyzopoulous *et al.* (1986) to study the rate of PO_4 sorption and release from Greek soils (Fig. 2.6). One sees a linear stage that appears Elovichian, but there is initially a section concave to the t axis. Polyzopoulous *et al.* (1987) ascribe this to a faster rate of PO_4 sorption that is non-Elovichian. One also sees a slow approach to equilibrium with time (Fig. 2.6).

Polyzopoulous *et al.* (1986) calculated t_0 values for the soils by extrapolating to the y-axes regression lines fitted to the linear Elovichian portion of the curves. The t_0 values calculated for the Dystrochrept and Palexeralf soils were greater than zero, which suggests that the assumption that t_0 is close to zero and can be neglected may not be valid.

Some investigators have used Elovich parameters to estimate reaction rates. Chien and Clayton (1980) suggested that a decrease in Y and/or an increase of X would increase reaction rate. However, this may be questionable. The slope of plots using an equation like Eq. (2.48) changes with the

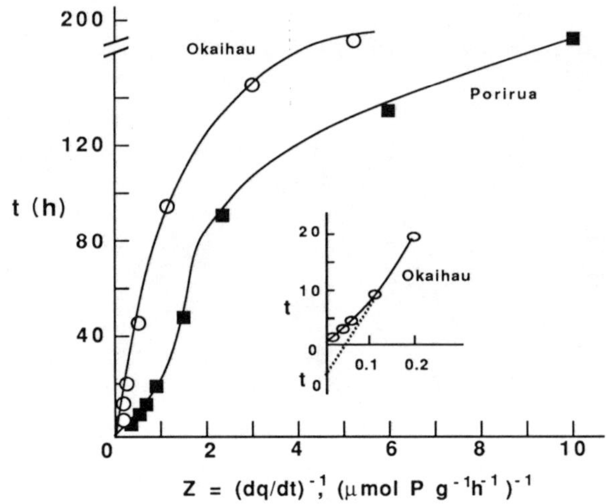

Figure 2.7. Plot of *t* versus *Z* for phosphate (PO_4) sorption on Porirua and Okaihau soils. [From Polyzopoulous *et al.* (1986)., with permission.]

level of added ion and with the solution to soil ratio (Sharpley, 1983). Consequently, these slopes are not always characteristic of the soil but depend on various experimental conditions. Sharpley (1983) used the Elovich equation to study soil phosphorus desorption and found that the X and Y values were related to the extractable Al content and $CaCO_3$ equivalent of acidic and basic soils, respectively.

Another criticism of using Eq. (2.48) is that the pre- and post-Elovichian sections are often not observed and one can erroneously conclude that the entire rate process is explainable using one kinetic law (Polyzopoulous *et al.*, 1986). When kinetic data are plotted according to Eq. (2.50) rather than Eq. (2.31) one sees pre- and post-Elovichian sections (Fig. 2.4 versus Fig. 2.7).

Some investigators have also suggested that "breaks" or multiple linear segments in a q versus ln t Elovich plot [Eq. (2.48)] could indicate a changeover from one type of binding site to another (Atkinson *et al.*, 1970; Chien and Clayton, 1980). However, one should be cautious about making such mechanistic conclusions from a plot of an empirical equation.

Parabolic Diffusion Equation

The parabolic diffusion law or equation can be used to determine whether diffusion-controlled phenomena are rate-limiting. This equation was originally derived based on radial diffusion in a cylinder where the ion concentration on the cylindrical surface is constant, and initially the ion concentration throughout the cylinder is uniform. It is also assumed that ion diffusion through the upper and lower faces of the cylinder (corresponding to external cleavage faces) is negligible. Following Crank (1975), if r is the radius of the cylinder, Q_t is the quantity of diffusing substance that has left the cylinder at time t, and Q_∞ is the corresponding quantity after infinite time, then

$$\frac{Q_t}{Q_\infty} = \frac{4}{\pi^{1/2}}\left(\frac{Dt}{r^2}\right)^{1/2} - \frac{Dt}{r^2} - \frac{1}{3\pi^{1/2}}\left(\frac{Dt}{r^2}\right)^{3/2} \tag{2.51}$$

For the relatively short times in most experiments, the third and subsequent terms may be ignored, and thus

$$\frac{Q_t}{Q_\infty} = \frac{4}{\pi^{1/2}}\left(\frac{Dt}{r^2}\right)^{1/2} - \frac{Dt}{r^2}$$

or

$$\frac{1}{t}\left(\frac{Q_t}{Q_\infty}\right) = \frac{4}{\pi^{1/2}}\left(\frac{D}{r^2}\right)^{1/2}\frac{1}{t^{1/2}} - \frac{D}{r^2} \tag{2.52}$$

and thus a plot of $\dfrac{(Q_t/Q_\infty)}{t}$ versus $1/t^{1/2}$ should give a straight line with a slope

$$\frac{4}{\pi^{1/2}}\left(\frac{D}{r^2}\right)^{1/2}$$

and intercept $(-D/r^2)$. Thus, if r is known, D may be calculated from both the slope and intercept.

A number of researchers have used the parabolic diffusion equation to study the kinetics of reactions on soil constituents (Chute and Quirk, 1967; Sivasubramaniam and Talibudeen, 1972; Evans and Jurinak, 1976; Vig *et al.*, 1979; Feigenbaum *et al.*, 1981; Sparks and Jardine, 1981; Jardine and Sparks, 1984; Havlin and Westfall, 1985; Hodges and Johnson, 1987); feldspar weathering (Wollast, 1967), and pesticide reactions (Weber and Gould, 1966). Sivasubramaniam and Talibudeen (1972) obtained parabolic plots for Al–K exchange on British soils that gave two distinct slopes, which the authors theorized could be indicative of two simultaneous diffusion-controlled reactions. They speculated that the rate-controlling step in Al^{3+} and K^+ adsorption was diffusion of the ions into the subsurface layers of the solid.

Power-Function Equation

Havlin and Westfall (1985) and Havlin *et al.* (1985) used a power-function equation to describe potassium release from soils. The integrated form of the power-function equation can be expressed (Havlin and Westfall, 1985) as

$$y = at^k \tag{2.53}$$

The linear transformation is

$$\ln y = \ln a + k(\ln t) \tag{2.54}$$

where y is the quantity of K released at time t, and a and k are constants. The k value is a rate-coefficient value.

Havlin and Westfall (1985) found that the power-function equation described potassium release from soils well. The a and k values were highly correlated with nonexchangeable potassium release. They found that k from Eq. (2.53) was highly correlated with potassium uptake and the relative yield of alfalfa (*Medicago sativa L.*) as shown in Fig. 2.8.

Comparison of Kinetic Equations

Comparisons of different kinetic equations for describing the kinetics of potassium reactions (Martin and Sparks, 1983; Sparks and Jardine, 1984; Havlin and Westfall, 1985), phosphorus sorption and release (Enfield *et al.*, 1976; Chien and Clayton, 1980; Onken and Matheson, 1982), sulfur sorption and desorption (Hodges and Johnson, 1987), and chromium, cadmium, and mercury retention/release (Amacher *et al.*, 1986) have appeared in the soil and environmental sciences literature. Some of these studies are summarized below.

Chien and Clayton (1980) compared several equations for describing PO_4 release from soils and found that the Elovich equation [Eq. (2.49)] was best based on the highest values of the simple correlation coefficient (r^2) and the lowest SE. The two-constant rate equation also described the data satisfactorily. The parabolic diffusion equation was judged unsatisfactory due to low r^2 and high SE values.

Onken and Matheson (1982) studied kinetics of phosphorus dissolution in EDTA (ethylenediamine tetraacetic acid) solution for several soils. They examined eight kinetic models (Table 2.2) and found that phosphorus dissolution in EDTA solution was best described using the two-constant rate, Elovich, and differential rate equations as indicated by high r^2 and low SE values. None of the models best described the dissolution for all soils.

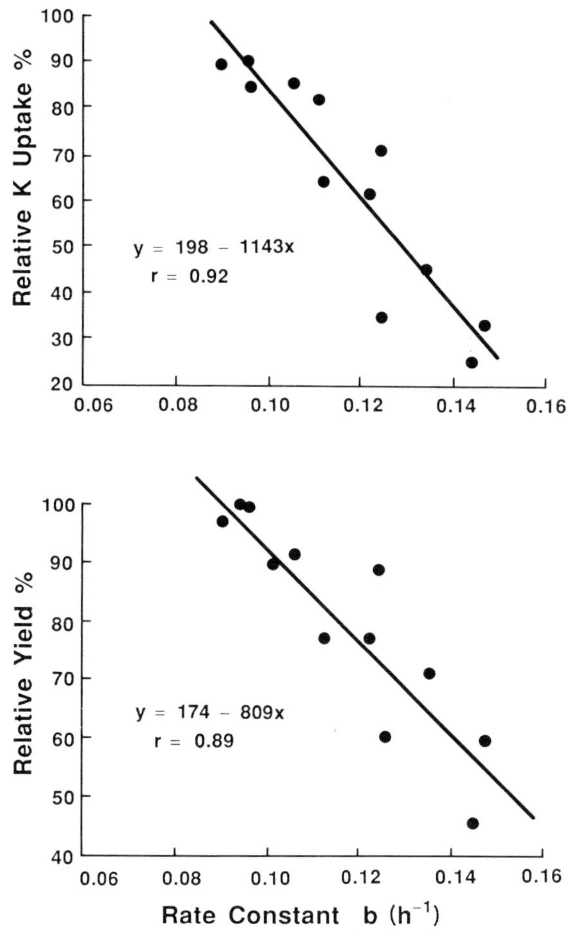

Figure 2.8. Relationship between relative potassium (K) uptake and yield in greenhouse experiment and potassium release constant k, as determined by Ca- resin extraction. [From Havlin and Westfall (1985), with permission.]

Sparks and Jardine (1984) found that the first-order equation best described potassium adsorption kinetics on clay minerals and soils. However, Havlin and Westfall (1985) reported that the power-function equation described nonexchangeable potassium release kinetics better than first-order or a number of other models.

Recently, Hodges and Johnson (1987) used five different kinetic equations to describe sulfur sorption and desorption on soils. Coefficients of determination showed that shell progressive particle diffusion, Elovich,

TABLE 2.2 Summary of r^2 and SE of Eight Kinetic Models for Phosphorus Dissolution in EDTA Solution from Six Test Locations Varying in Plant Response to Applied Phosphorus[a]

	Test number											
	1		7		14		15		16		20	
Kinetic model	r^2	SE	r^2	SE	r^2	SE	r^2	SE	r^2	SE	r^2	SE
Zero-order	0.96	0.04	0.68	0.26	0.87	0.45	0.66	0.30	0.70	0.83	0.86	0.74
First-order	0.92	0.05	0.58	0.34	0.76	0.20	0.55	0.34	0.56	0.39	0.72	0.99
Second-order	0.87	0.07	0.44	0.49	0.63	0.29	0.42	0.44	0.41	0.50	0.56	1.74
Third-order	0.82	0.12	0.41	0.84	0.51	1.23	0.30	2.70	0.30	1.35	0.44	3.06
Parabolic diffusion	0.95	0.05	0.86	0.17	0.97	0.07	0.81	0.22	0.85	0.24	0.96	0.36
Two-constant rate	0.87	0.05	0.93	0.15	0.98	0.05	0.88	0.19	0.91	0.18	0.99	0.26
Elovich-type	0.88	0.07	0.95	0.11	0.95	0.09	0.90	0.16	0.97	0.11	0.97	0.33
Differential rate	0.99	0.0001	0.91	0.0005	0.96	0.0003	0.93	0.001	0.97	0.0002	0.95	0.001

[a] From Onken and Matheson (1982), with permission.

and first-order equations described reactions well on the Cecil soil. But the statistical parameters were calculated for only the linear portion of the kinetic plots. When all data points were included, Hodges and Johnson (1987) report that the r^2 and SE values are altered significantly. The desorption data for sulfur reactions on the soils were best described by the shell progressive particle diffusion, Elovich, and parabolic diffusion equations (Hodges and Johnson, 1987).

The above studies clearly show that a number of different equations often describe rate data for soil constituents satisfactorily based on linear regression analyses. However, no single equation best describes every study, and conformity of data to a particular equation does not necessarily indicate that it is the *best* one to use. Moreover, one must be very careful not to attach mechanistic significance to linear plots based on the use of a given model.

TEMPERATURE EFFECTS ON RATES OF REACTION

Arrhenius and van't Hoff Equations

Increasing temperature usually causes a marked increase in reaction rate. Arrhenius observed the following relationship between k and temperature:

$$k = Ae^{-E/RT} \tag{2.55}$$

where k is the rate constant, A is a frequency factor, E is the energy of activation, R is the universal gas constant, and T is absolute temperature. Integrating Eq. (2.55) results in

$$\ln k = (\ln A) - E/RT \tag{2.56}$$

Thus, a plot of ln k versus $1/T$ would result in a linear relationship with the slope equal to $-E/R$ and the intercept ln A.

Energies of Activation. Low E values (<42 kJ mol^{-1}) usually indicate diffusion-controlled processes whereas higher E values indicate chemical reaction processes (Sparks, 1985, 1986). For example, E values of 6.7–26.4 kJ mol^{-1} were found for pesticide sorption on soils and soil components (Haque *et al.*, 1968; Leenheer and Ahlrichs, 1971; Khan, 1973) while gibbsite dissolution in acid solutions was characterized by E values ranging from 59 ± 4.3 to 67 ± 0.6 kJ mol^{-1} (Bloom and Erich, 1987).

Data taken from Huang *et al.* (1968) show the effect of temperature on the rate of potassium release from potassium-bearing minerals (Table 2.3).

TABLE 2.3 Apparent Rate Constants for the Release of Lattice Potassium from Potassium Minerals[a]

	Rate constant (h^{-1})	
	Temperature	
Mineral	301 K	311 K
Biotite	1.46×10^{-2}	3.09×10^{-4}
Phlogopite	9.01×10^{-4}	2.44×10^{-4}
Muscovite	1.39×10^{-4}	4.15×10^{-4}
Microcline	7.67×10^{-5}	2.63×10^{-4}

[a] From Huang *et al.* (1968), with permission.

A 10 K rise in temperature during the reaction period resulted in a two- to threefold increase in the rate constant.

One can also derive a relationship between temperature and the equilibrium constant K_{eq} and the standard free enthalpy $\Delta H°$. If the following reversible reaction is operational

$$\mathrm{A} \underset{k_d}{\overset{k_a}{\rightleftharpoons}} \mathrm{Y} \tag{2.57}$$

and by knowing

$$d(\ln K_{eq})/dT = \Delta H°/RT^2 \tag{2.58}$$

where $K_{eq} = (Y)/(A) = k_a/k_d$ and parentheses denote activity, the van't Hoff relationship can be written:

$$\frac{d(\ln k_a)}{dT} - \frac{d(\ln k_d)}{dT} = \frac{\Delta H°}{RT^2} \tag{2.59}$$

where

$$\frac{d(\ln k_a)}{dT} = \frac{E_a}{RT^2} \quad \text{and} \quad \frac{d(\ln k_d)}{dT} = \frac{E_d}{RT^2} \tag{2.60}$$

where k_a and k_d are adsorption and desorption rate coefficients, respectively and E_a and E_d are the energies of activation for adsorption and desorption, respectively.

Specific Studies

A number of researchers have studied the effect of temperature on reaction rates of soil chemical phenomena (Burns and Barber, 1961;

TABLE 2.4 Rate Constants for Indigenous Phosphorus Release from a Thiokol Silt Loam Soil[a]

	Surface soil: 0–0.30 m			Subsoil: 0.30–0.40 m		
	284 K	298 K	313 K	284 K	298 K	383 K
k_1 (s^{-1})	0.19×10^{-4}	0.51×10^{-4}	0.30×10^{-4}	0.24×10^{-4}	0.41×10^{-4}	0.41×10^{-4}
k_2 (s^{-1})	0.11×10^{-5}	0.27×10^{-5}	0.16×10^{-5}	0.15×10^{-5}	0.11×10^{-5}	0.26×10^{-5}
k_3 (s^{-1})	0.87×10^{-5}	0.13×10^{-6}	0.12×10^{-6}	0.15×10^{-6}	0.93×10^{-7}	0.19×10^{-6}

[a] The subscripts on k_1, k_2, and k_3 represent the first, second, and third phosphorus reactions, respectively. From Evans and Jurinak (1976), with permission.

Huang *et al.*, 1968; Griffin and Jurinak, 1974; Kuo and Lotse, 1974; Barrow and Shaw, 1975; Evans and Jurinak, 1976; Barrow, 1979; Sparks and Jardine, 1981; Ogwada and Sparks, 1986a, Hodges and Johnson, 1987). Evans and Jurinak (1976) studied the rate of phosphorus release from a Thiokol silt loam soil as a function of temperature, using strong anion exchange resins (Table 2.4). Phosphorus released from the surface and subsoil layers of the soil at 284, 298, and 313 K showed that during the initial 4 h of the reaction, the effect of temperature was small, with the rate of release increasing slightly as temperature increased. At times greater than 4 h, the effect of temperature was insignificant.

TRANSITION-STATE THEORY

Theory

Transition-state theory or reaction-rate theory was extensively developed by H. Eyring and collaborators (Glasstone *et al.*, 1941; Frost and Pearson, 1961).

For a given reaction in accordance with the absolute rate theory

$$A + B \underset{k_{-1}^{\ddagger}}{\overset{k_1^{\ddagger}}{\rightleftharpoons}} (AB)^{\ddagger} \xrightarrow{K} Y \tag{2.61}$$

where A and B are reactant molecules, $(AB)^{\ddagger}$ is an activated complex, $k_1^{\ddagger}$ is the rate of formation of the activated complex, $k_{-1}^{\ddagger}$ is the rate of decomposition of the activated complex, and K is the rate of product (Y) formation, with

$$K = K_t K^{\ddagger} \tag{2.62}$$

where K_t is the transmittance coefficient and $K^{\ddagger}$ is the pseudothermodynamic equilibrium constant of the activated complex, and

$$K_t = k_B T/h \tag{2.63}$$

where k_B is Boltzmann's constant and h is Planck's constant. Considering the laws of thermodynamics, the following expression may be developed:

$$K^{\ddagger} = e^{-\Delta G^{\ddagger}/RT} \tag{2.64}$$

where $\Delta G^{\ddagger}$ is the Gibbs energy of activation. Thus, the parameter, $k_1^{\ddagger}$, can be found using,

$$k_1^{\ddagger} = \left(\frac{k_B T}{h}\right) e^{-\Delta G_1^{\ddagger}/RT} \tag{2.65}$$

From the reaction rate theory (Frost and Pearson, 1961), it is found that

$$\Delta G_1^{\ddagger} = \Delta H_1^{\ddagger} - T\,\Delta S_1^{\ddagger} \tag{2.66}$$

where $\Delta H_1^{\ddagger}$ is the enthalpy of activation for the forward reaction and $\Delta S_1^{\ddagger}$ is the entropy of activation for the forward process.

Manipulation of Eq. (2.66) and substitution into Eq. (2.65) give

$$k_1^{\ddagger} = \frac{k_B T}{h} (e^{\Delta S_1^{\ddagger}/R - \Delta H_1^{\ddagger}/RT}) \tag{2.67}$$

Equation (2.67) would enable the calculation of $\Delta S_1^{\ddagger}$ because the following relationship for a unimolecular reaction is true:

$$\Delta H_1^{\ddagger} = E_1 - RT \tag{2.68}$$

where E_1 refers to an energy of activation for the forward reaction. The $\Delta G_1^{\ddagger}$ may be calculated using Eq. (2.66). Analogous expressions can be obtained for the reverse reaction through the use of E_{-1} and $k_{-1}^{\ddagger}$ calculated for the reverse process.

The pseudothermodynamic equilibrium constant of the activated complex ($K_{eq}^{\ddagger}$) is related to the thermodynamic state functions by

$$\Delta G^{\circ\ddagger} = -RT \ln K_{eq}^{\ddagger} = \Delta H^{\circ\ddagger} - T\,\Delta S^{\circ\ddagger} \tag{2.69}$$

where $\Delta G^{\circ\ddagger}$, $\Delta H^{\circ\ddagger}$, and $\Delta S^{\circ\ddagger}$ refer to the standard Gibbs energy of activation, the standard enthalpy of activation, and the standard entropy of activation, respectively.

In each case these parameters represent differences between the state function of the activated complex in a particular standard state and the state function of the reactants referred to in the same standard state. One is giving $K_{eq}^{\ddagger}$ all the characteristics of a thermodynamic equilibrium constant, although it should be multiplied by a transitional partition function. For ideal systems the magnitude of $\Delta H^{\circ\ddagger}$ does not depend on the choice of standard state, and for most of the nonideal systems that are encountered the dependence is slight. For all systems, the magnitudes of $\Delta G^{\circ\ddagger}$ and $\Delta S^{\circ\ddagger}$ depend strongly on the choice of standard state, so it is not useful to say that a particular reaction is characterized by specified numerical values of $\Delta G^{\circ\ddagger}$ and $\Delta S^{\circ\ddagger}$ unless the standard states associated with these values are clearly identified.

The reaction-rate theory assumes that colliding molecules (e.g., reactions between solution ions and ions held by an exchange complex) must be in a high energy state before a reaction can occur. This energy of activation is the result of van der Waals repulsive forces that occur as two ions approach each other. Without the repulsive forces, all exothermic reactions would have zero or very low activation energies and would be fast (Frost and Pearson, 1961). When these highly energized reactant molecules collide, they form an activated complex. It is a distinct chemical species in equilibrium with the reactants, which has gained one degree of vibrational freedom. This abnormal vibrational freedom causes the complex to lack a "restoring force," a force mandatory for all stable molecules: thus it flies apart in the period of one vibration (Denbigh, 1966). The activated complex is highly unstable and rapidly dissociates to form the eventual product or to reform the original reactants. Once the exchange reaction is complete, pseudothermodynamic parameters for the adsorption and the desorption process may be formulated.

A $\Delta G^{\ddagger}$ value may be considered as the free energy change between the activated complex and the reactants from which it was formed, all substances in reference to their standard states (Laidler, 1965). It is the $\Delta G^{\ddagger}$ value that determines the rate of the reaction (Glasstone *et al.*, 1941). The $\Delta G^{\ddagger}$ values should become larger with temperature, since the tendency of any reaction to proceed is hastened by temperature increases.

The enthalpy of activation $\Delta H^{\ddagger}$ is a measure of the energy barrier that must be overcome by reacting molecules (Frost and Pearson, 1961). The energy needed to raise the molecules from their ground state to one of an excited state is the sum of the electronic, vibrational, rotational, and translational energy terms. Thus, the energy needed to change the orientation, structure, and position of an ion from one phase to that of another is the total heat energy required in the process. Variations in $\Delta H^{\ddagger}$ with temperature are not the result of energy changes involved in making or breaking bonds, but rather are due to alterations in the heat-capacity behavior of the ions involved in the exchange reaction (Frost and Pearson, 1961).

The entropy of activation $\Delta S^{\ddagger}$ may be regarded as the "saddle point of energy" over which reactant molecules must pass as activated complexes (Frost and Pearson, 1961; Laidler, 1965). The $\Delta S^{\ddagger}$ conveys whether a particular reaction proceeds more quickly or slowly than another individual reaction. Negative $\Delta S^{\ddagger}$ values would depict a system that could ascertain a more ordered molecular arrangement in a shorter period of time relative to a positive or less negative $\Delta S^{\ddagger}$ parameter. The molecular arrangement would be related to both the aqueous and solid phases where ion hydration and configurational entropy constituents are considered.

Application to Soil Constituent Systems

Griffin and Jurinak (1974) calculated pseudothermodynamic parameters for phosphate interactions with calcite using reaction-rate theory. Gonzalez *et al.* (1982) applied reaction-rate theory to a treatment of adsorption–desorption processes on an Fe–selica gel system. In 1981, Sparks and Jardine applied reaction-rate theory to kinetics of potassium adsorption and desorption in soil systems for the first time (Table 2.5).

In their systems, the $\Delta G^{\ddagger}$ values were higher for desorption than for adsorption, suggesting a greater free-energy requirement for potassium desorption. The $\Delta G^{\ddagger}$ values for both adsorption and desorption were also slightly higher in the B21t than in the Ap soil horizon, suggesting slower

TABLE 2.5 Kinetic Parameters for Potassium Adsorption and Desorption Processes at Three Temperatures in Matapeake Ap and B21t Horizons Using Reaction-Rate Theory[a]

Temperature (K)	$\Delta G^{\ddagger}$ (kJ mol^{-1})	$\Delta H^{\ddagger}$ (kJ mol^{-1})
Ap horizon		
Adsorption		
276	67.29	13.74
298	71.44	13.58
313	74.50	13.45
Desorption		
276	72.74	20.83
298	76.85	20.62
313	79.63	20.53
B21t horizon		
Adsorption		
276	68.21	16.30
298	72.28	14.88
313	75.46	14.75
Desorption		
276	73.20	20.66
298	77.10	20.49
313	80.32	20.36

[a] From Sparks and Jardine (1981), with permission.

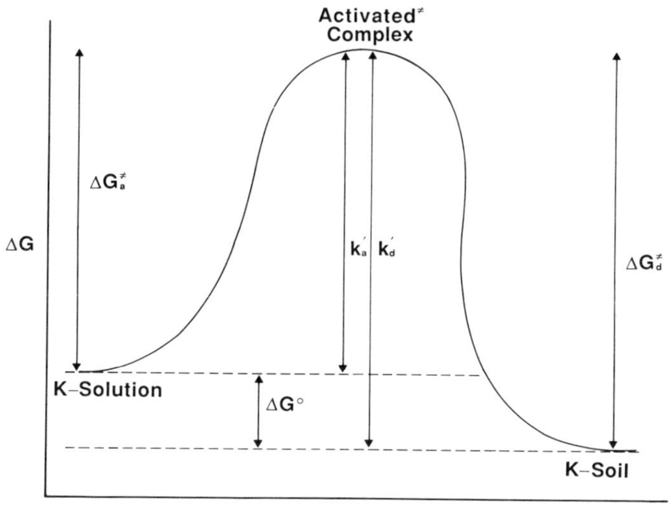

Figure 2.9. Schematic diagram of ΔG versus reaction coordinate for potassium exchange on a Matapeake soil where ΔG is the Gibbs free energy. [From Sparks and Jardine (1981), with permission.]

reactions due to more restrictive binding sites for K in the B21t horizon (Glasstone *et al.*, 1941). Figure 2.9 illustrates a schematic correlation between the pseudothermodynamic parameters and those established using thermodynamics of ion exchange theory. The $\Delta G_a^\ddagger$ is the change in free energy required for potassium to cross the barrier of adsorption at an apparent rate of k'_a. The $\Delta G_d^\ddagger$ represents the change in free energy needed by the reverse reaction of desorption at the apparent rate of k'_d. The difference between these two parameters yeilds $\Delta G°$, the Standard Gibbs free energy.

The $\Delta H^\ddagger$ values in both horizons were higher for desorption than for adsorption (Table 2.5), suggesting that the heat energy required to overcome the potassium desorption barrier was greater than that for potassium adsorption. This was also seen in the magnitude of the E_a and E_d values calculated (not shown). A schematic correlation between $\Delta H°$ *and* $\Delta H^\ddagger$ can be observed like that shown in Fig. 2.9 for $\Delta G°$ and $\Delta G^\ddagger$. The $\Delta H_a^\ddagger$ represents the change in heat energy needed for K^+ to go from the solution phase to the solid phase (adsorption), whereas $\Delta H_d^\ddagger$ is the heat-energy requirement for the desorption reaction. The difference in these two parameters represents $\Delta H°$ (Frost and Pearson, 1961).

SUPPLEMENTARY READING

Aharoni, C., and Ungarish, M. (1976). Kinetics of activated chemisorption. I. The non-Elovichian part of the isotherm. *J. Chem. Soc., Faraday Trans.* **72,** 400–408.

Bunnett, J. F. (1986). Kinetics in solution. *In* "Investigations of Rates and Mechanisms of Reactions" (C. F. Bernasconi, ed.), 4th ed., pp. 171–250. Wiley, New York.

Chien, S. H., and Clayton, W. R. (1980). Application of Elovich equation to the kinetics of phosphate release and sorption in soils. *Soil Sci. Soc. Am. J.* **44,** 265–268.

Crank, J. (1975). "The Mathematics of Diffusion." Oxford Univ. Press (Clarendon), London and New York.

Denbigh, K. G. (1966). "The Principles of Chemical Equilibrium with Applications in Chemistry and Chemical Engineering." Cambridge Univ. Press, London and New York.

Eyring, H., Lin, S. H., and Lin, S. M. (1980). "Basic Chemical Kinetics." Wiley, New York.

Froment, G. F., and Bischoff, K. B. (1979). "Chemical Reactor Analysis and Design." Wiley, New York.

Frost, A. A. and Pearson, R. G. (1961). "Kinetics and Mechanisms." Wiley, New York.

Gardiner, W. C. Jr. (1969). "Rates and Mechanisms of Chemical Reactions." Benjamin, New York.

Glasstone, S., Laidler, K. J., and Eyring, H. (1941). "The Theory of Rate Processes." McGraw-Hill, New York.

Hammes, G. G. (1978). "Principles of Chemical Kinetics." Academic Press, New York.

Laidler, K. J. (1965). "Chemical Kinetics." McGraw-Hill, New York.

Levenspiel, O. (1972). "Chemical Reaction Engineering." Wiley, New York.

Moore, J. W., and Pearson, R. G. (1981). "Kinetics and Mechanism." Wiley, New York.

Polyzopoulos, N. A., Keramidas, V. Z., and Pavatou, A. (1986). On the limitations of the simplified Elovich equation in describing the kinetics of phosphate sorption and release from soils. *J. Soil Sci.* **37,** 81–87.

Skopp, J. (1986). Analysis of time dependent chemical processes in soils. *J. Environ. Qual.* **15,** 205–213.

Sparks, D. L. (1965). Kinetics of ionic reactions in clay minerals and soils. *Adv. Agron.* **38,** 231–266.

Sparks, D. L. (1986). Kinetics of reactions in pure and in mixed systems. *In* "Soil Physical Chemistry" (D. L. Sparks, ed.), pp. 83–178. CRC Press, Boca Raton, Florida.

Sparks, D. L. (1987). Kinetics of soil chemical processes: Past progress and future needs. *In* "Future Developments in Soil Science Research" (L. L. Boersma *et al.,* eds.), pp. 61–73. Soil Sci. Soc. Am., Madison, Wisconsin.

3

Kinetic Methodologies and Data Interpretation for Diffusion-Controlled Reactions

INTRODUCTION

One of the most important aspects of a kinetic study on soil constituents is the method one uses to measure rate coefficients and other kinetic parameters. At the outset, it should be realized that no kinetic method currently available is perfect. Each has its own advantages and disadvantages, and these must be assessed carefully before using.

Additionally, the types of reactions that are studied and their relative time scales are important considerations in choosing a kinetic method. For example, reactions that are exceedingly rapid and occur on microsecond, millisecond, and second time scales cannot be measured using traditional batch and flow techniques. However, some weathering and pesticide reactions in soils and on soil constituents, are slow, diffusion-controlled processes. In these systems, certain batch and flow methods would be quite satisfactory.

Another consideration in choosing a kinetic method is the objective of one's experiments. For example, if chemical kinetics rate constants are to be measured, most batch and flow techniques would be unsatisfactory since they primarily measure transport- and diffusion-controlled processes, and apparent rate laws and rate coefficients are determined. Instead, one should employ a fast kinetic method such as pressure-jump relaxation, electric field pulse, or stopped flow (Chapter 4).

Regardless of the kinetic method one chooses, controlling the temperature is imperative. Because most reaction rates are strongly temperature-dependent, it is necessary that the temperature be maintained at a constant level for any given experiment and that it be known (Bunnett, 1986). This can easily be obtained using constant temperature baths or temperature-controlled incubators and chambers.

The objectives of this chapter are to discuss kinetic methods that are available for studying reactions on soil constituents, how one can analyze the data from these techniques, and the advantages and disadvantages of each method.

HISTORICAL PERSPECTIVE

Two primary methods have been employed to measure reaction rates on soil constituents—batch and flow. Thompson (1850) and Way (1850) conducted cation exchange experiments using columns of soil through which adsorptive solutions were leached. As the adsorptive solution passed through the soil it miscibly displaced the existing solution, allowing the incoming cations to displace the existing cations on the colloid. The displaced cations were removed from the reaction site as the leachate was collected. The preceding experiments were likely the first examples of miscible displacement techniques.

Way's work was caustically criticized by the celebrated chemist Liebig (Thomas, 1977). Liebig believed that Way's "exchange" was simply caused by cations being held within the capillaries of the soil column, much like water in a sponge. If true, this implied that the length and packing of the column would affect the exchange capacity of the adsorbent. The realization that columns of the same soil did not always yield similar results led soil scientists, including Way, to use digestion procedures (Kelley, 1948). These methods involved placing known quantities of adsorbent and the adsorptive in a closed vessel, and after time (digestion), analyzing the liquid. Thus, this was the beginning of the batch technique for studying the kinetics of reactions. However, the above techniques were quite cumbersome, and consequently did not enable soil scientists to accurately study the rates of the exchange reactions.

BATCH TECHNIQUES

Advantages and Disadvantages

Many kinetic studies investigating reactions on soil constituents have used batch techniques. The traditional batch or tube technique involves placing an adsorbent and the adsorptive in a vessel such as a centrifuge tube. The suspension is stirred or agitated using a reciprocating shaker. Then the suspension is usually centrifuged or filtered to separate a clear supernatant solution for subsequent analysis.

The use of centrifugation to separate the liquid from solid phases in traditional batch or tube techniques has several disadvantages. Centrifugation could create electrokinetic effects close to soil constituent surfaces that would alter the ion distribution (van Olphen, 1977). Additionally, unless filtration is used, centrifugation may require up to 5 min to separate the solid from the liquid phases. Many reactions on soil constituents are complete by this time or less (Harter and Lehmann, 1983; Jardine and Sparks, 1984; Sparks, 1985). For example, many ion exchange reactions on organic matter and clay minerals are complete after a few minutes, or even seconds (Sparks, 1986). Moreover, some reactions involving metal adsorption on oxides are too rapid to be observed with any batch or, for that matter, flow technique. For these reactions, one must employ one of the rapid kinetic techniques discussed in Chapter 4.

Also, to measure properly the kinetics of a reaction, the technique should not alter the reactant concentration significantly (Zasoski and Burau, 1978). Thus, the sample and the suspension should have a similar solid to solution ratio at all times. Unfortunately, this has not been the case in most batch studies (Barrow, 1983). Most kinetic batch studies involving soil constituents have used large solution: soil ratios where the concentration in the solution and the quantity of adsorption vary simultaneously.

Exceptions are techniques employed by Zasoski and Burau (1978), van Riemsdijk (1970), and van Riemsdijk and de Haan (1981). In these studies, the solution concentration was held constant. Unless batch techniques similar to those cited above are used, these conditions do not apply to experiments in which a wide solution:soil ratio is used. These techniques will be elaborated on later in this chapter.

Another problem with the batch technique is mixing the adsorptive and adsorbent. If mixing is inadequate, the rate of reaction is limited and significant mass transfer exists. However, vigorous mixing, which many investigators have used, can cause abrasion of the soil constituent particles, leading to high rates of reaction and even changes in the surface chemistry of the particles (Barrow and Shaw, 1979; Ogwada and Sparks, 1986b, c). The abrasion of particles could be a serious problem, particularly when elements, like potassium, are being studied that are contained within the particle structure (Sparks, 1985, 1986).

In many batch and flow methods, apparent rate laws are measured since the degree of mixing (shaking, stirring, flowing) affects rates of reaction (Ogwada and Sparks, 1986b). Consequently, mixing does not totally eliminate diffusion. Film diffusion (diffusion of the adsorptive through an imperfectly mixed layer or film around the particle) may be greatly reduced by mixing, but unless extreme agitation is employed, particle diffusion and other mass transfer phenomena cannot be eliminated.

With all batch techniques, there is the common problem of not removing the desorbed species. This can cause an inhibition of further adsorbate release (Sparks, 1985, 1987a), promote hysteretic reactions, and create secondary precipitation during dissolution of soil minerals (Chou and Wollast, 1984). However, one can use either exchange resins or sodium tetraphenylboron, which is quite specific for precipitating released potassium, as sinks for desorbed species and still employ a batch technique (Sparks, 1986). Also, since in most batch methods the reverse reactions are not controlled, problems are created in calculating rate coefficients. This is particularly true for heterogeneous systems such as soils.

Specific Batch Techniques

Many of the disadvantages of batch techniques just given can be eliminated by using a batch technique developed by Zasoski and Burau (1978) to study the rate of metal sorption on colloids and successfully used by Harter and Lehmann (1983) to study metal reaction kinetics on soils. A schematic diagram of the apparatus used in this technique is shown in Fig. 3.1.

With this apparatus, constant pH can be maintained by using a combination glass electrode along with an automatic titrimeter and digital buret.

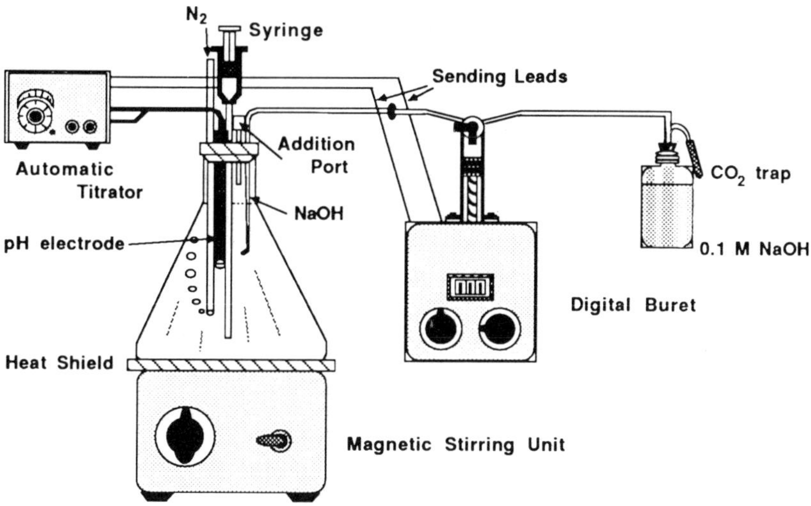

Figure 3.1. Schematic diagram of equipment used in batch technique of Zasoski and Burau (1978). (Reprinted with permission of the publisher).

The ease with which one can maintain constant pH is a major attribute of this apparatus, since many sorption reactions on soil constituents are affected by pH. One can also exclude oxygen and prevent oxidation of metals and CO_2 production by using N_2.

With this method, an adsorbent is placed in a vessel containing the adsorptive, pH and suspension volume are adjusted, and the suspension is vigorously mixed using a magnetic stirrer. At various times, suspension aliquots are withdrawn using a syringe containing N_2 gas to prevent CO_2 and O_2 from entering the reaction vessel. The suspension is quickly filtered using a membrane filter (Fig. 3.1) and the filtrates are then weighed and analyzed.

This technique does not use centrifugation to obtain a clear supernatant solution. Sampling takes about 10 s, and filtration 2–3 s. Thus, an advantage of this technique is that one could observe reactions at 15-s intervals. Additionally, excellent mixing occurs within the reaction vessel, and one can maintain a constant solid-to-solution ratio throughout an experiment (Table 3.1). Other data presented by Zasoski and Burau (1978) showed that repeated sampling did not much change the solid-to-suspension ratio or affect the reaction parameters.

Variations of the Zasoski and Burau (1978) technique have been employed by van Riemsdijk and de Haan (1981) and Phelan and Mattigod (1987). van Riemsdijk and de Haan (1981) investigated PO_4 sorption

TABLE 3.1 Solid-to-Solution Ratio of Samples Taken during a Kinetic Run[a]

Time (min)	Sample volume (ml)	Weight of MnO_2 (mg)	Ratio ($ml\ mg^{-1}$)
0	21.22	2.15	9.86
0.5	10.71	0.85	12.60
1.0	11.47	1.07	10.71
2.0	11.58	1.30	8.90
3.0	10.90	1.21	9.00
4.0	11.43	1.30	8.79
5.0	11.75	1.20	9.79
			$9.81 = \bar{X}$

[a] Expected ratio is 10 ml mg^{-1} MnO_2. From Zasoski and Burau (1978), with permission.

kinetics on soil at constant supersaturation with respect to metal phosphates using a phosphato-stat (c_p-stat) technique. This method allowed one to accurately measure reaction rates at constant supersaturation.

Phelan and Mattigod (1987) studied calcium phosphate precipitation from stable supersaturated solutions using pH/Ca-stat and pH-stat. The pH and Ca^{2+} activity of the titrand were kept constant utilizing ion-specific electrodes attached to automatic titrators. A schematic diagram of the apparatus used by Phelan and Mattigod is shown in Fig. 3.2.

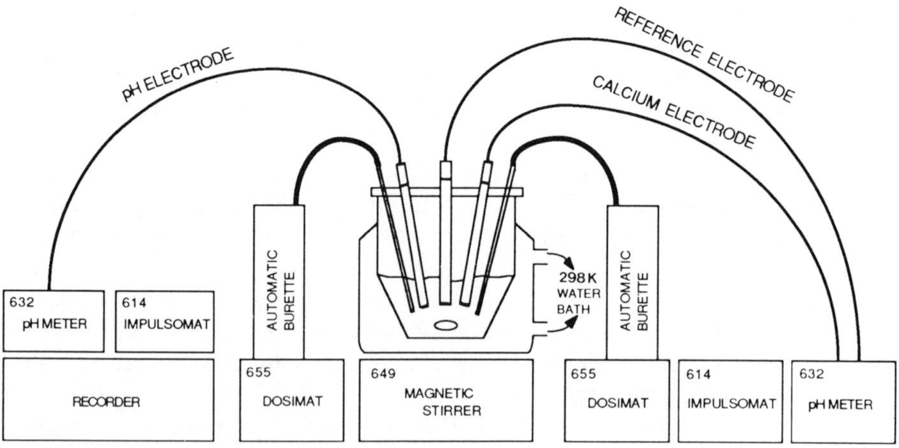

Figure 3.2. Schematic diagram of apparatus used in precipitation experiments of Phelan and Mattigod (1987). (Reprinted with permission of the publisher.)

There are several disadvantages to the batch techniques of Zasoski and Burau (1978), van Riemsdijk and de Haan (1981), and Phelan and Mattigod (1987). A major disadvantage is in not removing desorbed species and difficulty in monitoring desorption kinetics. In studies where desorbed species in the solution phase inhibit further release of adsorbate, such as potassium, or could cause secondary precipitation reactions, these methods may not be suitable.

Another potential problem with the batch method of Zasoski and Burau (1978) is keeping the soil or colloid uniformly suspended. This could be difficult with sandy soils where the sand-sized particles could sink to the bottom of the reaction vessel, or with high organic matter soils. With the latter, humic materials could rise to the surface of the reaction vessel creating a nonuniform suspension.

Although mixing by stirring is used in all the above batch methods and is desirable to diminish mass-transfer phenomena, the effect of mixing on the surface area of the adsorbent with reaction time should be assessed with each of these methods, or with any kinetic technique. Stirring the reaction system may cause some degradation of colloidal particles (Barrow and Shaw, 1979; Ogwada and Sparks, 1986b), consequently increasing the

TABLE 3.2 Effect of Degree of Agitation on Surface Area and Rate of K^+ Adsorption on Chester Loam[a]

Mixing rates (rpm)	Specific surface ($\times 10^4$ m^2 kg^{-1})	Rate coefficient k_a (min^{-1})
Stirred		
0	5.35	0.027
285	5.37	0.048
330	5.37	0.081
370	5.37	0.251
435	5.37	0.250
478	5.37	0.251
640	5.54	0.315
670	6.11	0.341
Vortex batch		
2240	5.37	2.330
2290	5.38	2.330
2318	5.38	2.329
2420	5.54	3.501
2490	5.56	3.512
2625	5.69	3.610
2700	5.69	3.611

[a] From Ogwada and Sparks (1986c), with permission.

surface area of the sorbent during the experiment. This could result in an increased rate of reaction with time.

For example, Ogwada and Sparks (1986c) found that specific surface for a soil was relatively constant under stirred conditions for mixing rates of 0–478 rpm, but increased abruptly at higher mixing rates. With vigorous vortex mixing, specific surface increased above a 2318-rpm mixing rate (Table 3.2). Such conditions would be very undesirable with any kinetic method.

Data Analysis

With batch techniques, the amount of sorbed species q in mol kg^{-1} may be calculated using the following equation:

$$\frac{C_f V_f - C_o V_o}{m} = q \tag{3.1}$$

where C_f and C_o are final and initial sorptive concentrations, respectively, in mol/liter, V_f and V_o are the final and initial sorptive volumes, respectively, in liters, and m is the mass of the sorbent in kilograms.

FLOW AND STIRRED-FLOW METHODS

Advantages and Disadvantages

Flow methods have recently been used in a number of kinetic studies of soils and soil constituents (Sivasubramaniam and Talibudeen, 1972; Sparks *et al.*, 1980b; Chou and Wollast, 1984; Carski and Sparks, 1985; Miller *et al.*, 1986; Hodges and Johnson, 1987; Schnabel and Fitting, 1988). These techniques, like batch methods, are not a panacea for kinetics analyses; there are advantags and disadvantages.

Sparks *et al.* (1980b) introduced a continuous flow method (next subsection) that is quite similar in principle to liquid-phase column chromatography. This method was used to study potassium adsorption dynamics on soils and clay minerals (Sparks and Jardine, 1981; Sparks and Rechcigl, 1982; Jardine and Sparks, 1984; Ogwada and Sparks, 1986a,b,c), silicate sorption on soils (Miller *et al.*, 1986), SO_4 sorption and desorption on soils (Hodges and Johnson, 1987), and Al reactions on clay minerals and peat (Jardine *et al.*, 1985a).

Using this technique, one can measure reactions at rapid intervals (~1 min). This method may also be preferable if one wishes to simulate solute transport in soils. Most batch techniques, as discussed earlier, have employed high solution : solid ratios that change throughout the reaction study. However, flow techniques like that of Sparks *et al.* (1980b) employ small solution : solid ratios (usually ≤ 1), which are constant throughout an investigation.

The amount of solution in contact with colloidal particles is also an important attribute of a flow technique. Supplied with a solution of the same concentration, soil constituent particles with solution flowing past them will be exposed to a greater mass of adsorptive (concentration $\times$ flow rate $\times$ time) than the particles in a static system (concentration $\times$ solution volume) by the time an equilibrium is attained. More important, desorbed species that were originally on the adsorbent are constantly being removed (Akratanakul *et al.*, 1983; Sparks, 1985, 1986).

With a closed system like a batch technique, replacement of adsorbate species cannot be complete unless the concentration of the adsorptive is increased in bulk solution. This forces the adsorptive back onto the adsorbent. In an open flow system, ion exchange, for example, can be complete, since the replaced adsorbate ions are always removed from the system and more of the introduced adsorptive is added (Akratanakul *et al.*, 1983).

There are, however, a number of disadvantages to using continuous flow techniques to study the kinetics of reactions on soil constituents. Often the colloidal particles are not dispersed—for example, the time required for an adsorptive to travel through a thin layer of collodial particles is not equivalent at all locations of the layer. Consequently, mass transfer can be significant if the sample is not dispersed. Skopp and McAllister (1986) note that even if the sample is dispersed, different pore and particle sizes of the adsorbent may result in "nonuniform tracer transit times." The thickness of the disc of colloidal particles should be thin; and measured to establish that perfect mixing is operational.

With a continuous flow method, flowing is the sole way the sample is mixed. Consequently, there may be imperfect mixing. Thus, the concentration of the adsorptive in the flow chamber may not equal the effluent concentration; this is because transport and chemical kinetics phenomena are both occurring simultaneously.

Additionally, the lack of adequate mixing with a continuous flow method results in pronounced diffusion (Ogwada and Sparks, 1986a,b) and the determination of apparent rate parameters.

Thus, it is important to remember that with many kinetic techniques that are currently used to study reactions on soil constituents one is usually measuring diffusion-controlled kinetics. Certainly, this fact does not diminish the importance of such investigations, but rather emphasizes that *kinetic* events are being studied rather than *chemical kinetics* (Chapter 2).

Continuous Flow Method

The method developed by Sparks *et al.* (1980b) is a good example of a continuous flow and is shown in Fig. 3.3. Samples can either be injected as suspensions or spread as dry samples on a membrane filter. The filter holder is capped securely and then is attached to a fraction collector and a peristaltic pump, which maintains a constant flow rate. Samples are leached with sorptive solutions, and effluents are collected at various time intervals.

For sorption/desorption studies, the sorption reaction is followed by monitoring the increasing concentration of leachate with time. At an apparent equilibrium, the effluent concentration equals that of the initial sorptive solution. The desorption reaction is studied in a similar way, the reaction being followed by monitoring the decreasing concentration of the previously sorbed ion or other sorbate. In either case, the reaction is followed by determining the sorptive concentration in solution. This means that any process that effects a change in concentration will be interpreted as adsorption or desorption.

With the continuous flow method of Sparks *et al.* (1980b), the dilution of incoming sorptive solution by the liquid used to load the sorbent onto the

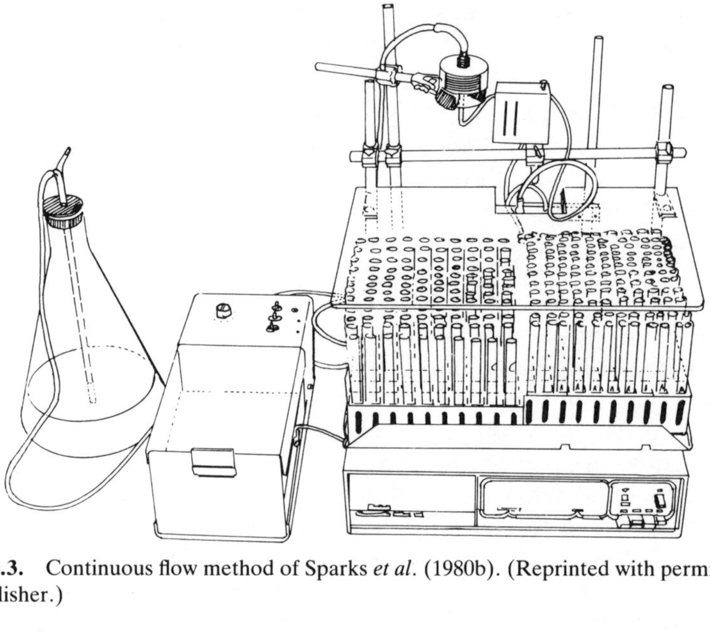

Figure 3.3. Continuous flow method of Sparks *et al.* (1980b). (Reprinted with permission of the publisher.)

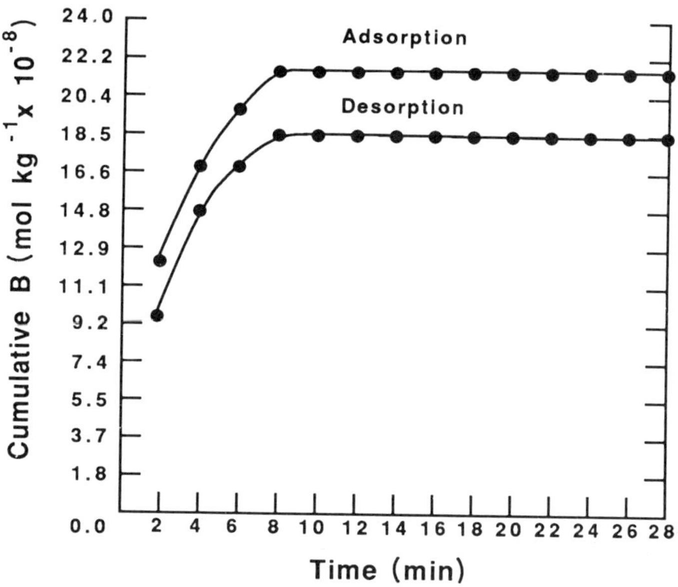

Figure 3.4. Apparent adsorption and desorption of boron (B), determined using the continuous flow technique (Sparks *et al.*, 1980b) with no adsorbent. [From Carski and Sparks (1985), with permission.]

filter (if the sorbent is added to the filter as a suspension), or the washing out of leftover sorptive solution during desorption can result in concentration changes not due to sorption or desorption. These concentration changes are potential sources of error, as shown by Carski and Sparks (1985). In Fig. 3.4 one sees that with acid-washed sand and no sorbent where no B sorption or desorption would be expected, the continuous flow technique of Sparks *et al.* (1980b) predicts both to occur. In this instance, the dilution effect or washing-out effect alone may account for the total amounts "sorbed" or "desorbed." Regardless of the sorbent used, some solution will be held back or entrained on top of the filter. This entrained solution cannot be completely removed by suction, and the total amount will be dependent on the water-holding capacity of the sorbent. Moreover, the dependence on the amount of entrained fluid on the nature of sorbent means that the magnitude of the dilution effect will be different for each sorbent. Fortunately, the dilution problem and other shortcomings of continuous flow techniques discussed earlier can be eliminated by using a stirred-flow method (Carski and Sparks, 1985), discussed later.

Fluidized Bed Reactors

The fluidized bed reactor has been used by engineers, chemists, and geochemists to study various kinetic phenomena (Chou and Wollast, 1984; Holdren and Speyer, 1985, 1987; Wollast and Chou, 1985). It is widely used in the chemical industry for physical and chemical processes involving a solid phase and a gas or liquid phase. The flow of the fluid is adjusted such that its velocity equals the settling rate of the solid particles in the particular suspension. Because the suspension is often quite dense, the settling rates of different-sized particles are equalized by frequent collisions with other particles. It is best to utilize a well-defined size fraction. This is especially true if one is using a small reactor to study the kinetics of reactions. If a particle escapes from the fluidized bed to the overlying fluid, it is then placed in a medium of lower density and then falls quickly back into the fluidized bed. It is therefore possible to obtain over a given height, depending on the flow rate of the fluid, a homogeneous suspension where the solid and the fluid are rapidly mixed (Chou and Wollast, 1984). For more detail on the theory of fluidized bed reactors the reader can consult most chemical engineering texts (e.g., Zenz and Othmer, 1960).

Chou and Wollast (1984) used a fluidized bed reactor to study albite weathering. An illustration of their device is shown in Fig. 3.5. The flow needed to maintain the feldspar particles in suspension is provided by the pumping rate P_1, while P_2 is the rate of addition of fresh solution; P_2 is also the rate of output of the reacted solution. By changing the rate of renewal of P_2 one can vary the residence time of the fluid in the reactor. To maintain a small difference in concentration between the input at the bottom of the fluidized bed and the output at the top of the bed, P_2 must be small in comparison to P_1. Chou and Wollast (1984) maintained the renewal rate P_2 between 3 and 6% of the mixing rate P_1.

Using the fluidized bed reactor of Chou and Wollast (1984), the rate of reaction is obtained by multiplying the renewal rate by the difference in concentration between the input and the output solutions. The rate is then normalized in relation to the total surface area of the solid.

The fluidized bed reactor has been used by several researchers to study the kinetics of chemical weathering (Holdren and Speyer, 1985, 1987; Chou and Wollast, 1985). One of the advantages in using the fluidized bed reactors for studies of this type is that there are no strong concentration gradients in the aqueous and solid phases. Additionally, the concentration of the dissolved species can be maintained at levels well below saturation with respect to possible precipitates. This means, for example, that one could study mineral dissolution exclusively without secondary precipita-

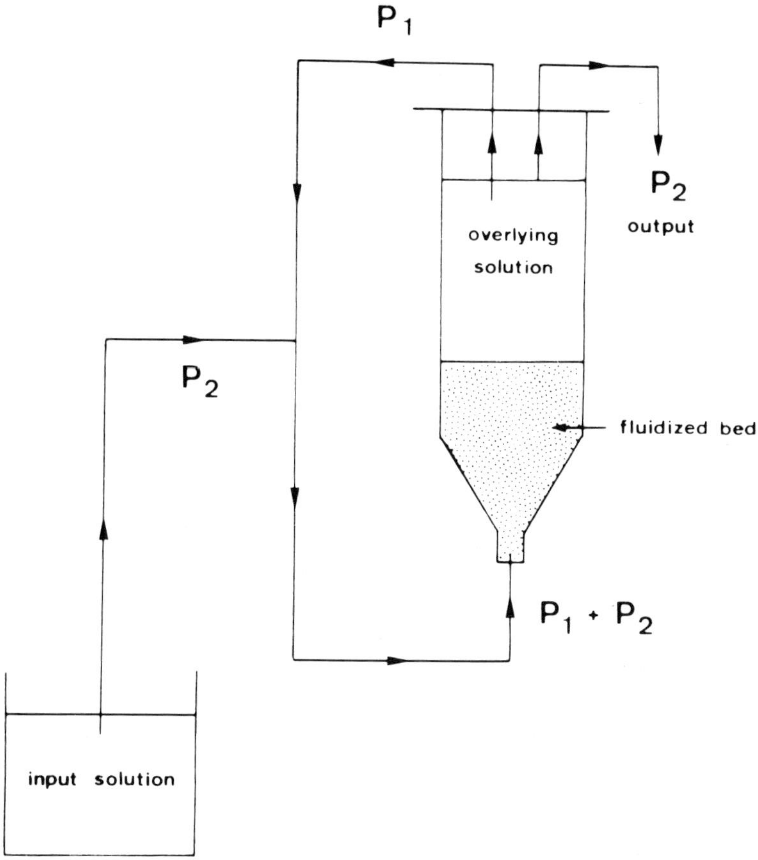

Figure 3.5. Schematic representation of the fluidized bed reactor. P_1 is the rate required to keep the particles in suspension. P_2 is the rate of addition of fresh input solution. [From Chou and Wollast (1984), with permission.]

tion interference. Moreover, one can easily evaluate the effect of various chemical conditions on the dissolution rate of the same sample of solid by abruptly changing the composition of the input solution without manipulating the solid phase.

Stirred-Flow Technique:

Stirred-flow reactors have been studied and used by chemical engineers for many years, but their application to chemical research is more recent, first by Denbigh (1944), and then by Hammett (1960). Stirred-flow reactors have recently been used by soil chemists to study soil chemical reaction rates (Carski and Sparks, 1985; Randle and Hartmann, 1987; Seyfried *et al.*, 1988).

Basic Design and Characteristics. The stirred-flow technique developed by Carski and Sparks (1985) is shown in Fig. 3.6. The basic components for the construction of this device include the barrel and plunger from a 30-ml plastic syringe and a 25-mm Nuclepore Swin Lok filter holder. The filter holder is modified and glued to the top of the syringe barrel, and the base

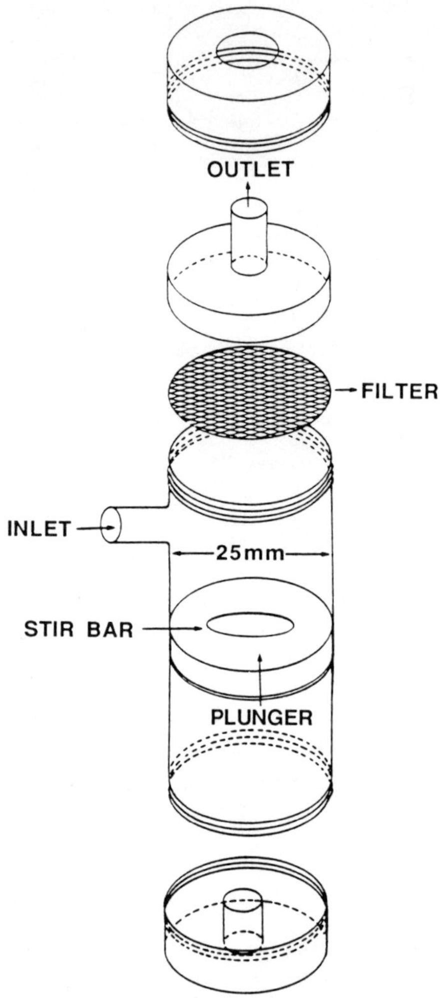

Figure 3.6. Schematic diagram of stirred-flow reaction chamber. [From Carski and Sparks (1985), with permission.]

of the barrel is threaded to provide plunger height adjustment. The device enables one to add and to maintain a known quantity of fluid to a known amount of sorbent regardless of the sorbent used. The added fluid represents the entrained fluid on the continuous flow filter. A magnetic stirring bar is placed in the chamber above the plunger, a known amount of dry sorbent is loaded into the reaction chamber, a 0.2-μm membrance filter and the top are attached, and a known volume of entrained fluid is added using a hypodermic syringe. A plunger is then used to displace the excess air from within the reaction chamber, thus enabling a known volume to be diluted or washed out. This volume is maintained throughout the sorption and desorption reactions. A peristaltic pump is used to maintain a constant flow rate and a fraction collector is used to collect leachates. A magnetic stirrer is used to ensure perfect mixing in the reaction chamber. In the original method of Carski and Sparks (1985), stirring speed was kept to about 100 rpm to minimize abrasion of the sorbent.

The stirred-flow technique is an improvement over the continuous flow method described earlier. The method effects perfect mixing (Seyfried *et al.*, 1988) so that the chamber and effluent concentrations are euqal and transport phenomena are minimized significantly. Additionally, the sorbent is dispersed and the dilution error present in the continuous-flow method can be accounted for. The stirred-flow technique also retains the attractive features of removing desorbed species at each step of the reaction process and of easily studying desorption kinetics phenomena.

Data Analyses Using Continuous Flow and Stirred-Flow Methods

Since continuous flow and stirred-flow methods include a physical parameter, flow rate, data analyses used for batch techniques are inappropriate.

To analyze data using these two methods one must make two assumptions: (1) that a sorptive entering the chamber can either be sorbed or remain in solution, and (2) the sample is perfectly mixed i.e., the concentration in the mixing chamber equals the effluent concentrations. With these assumptions, one can then develop an equation for mass balance which can be used to analyze time-dependent data using a continuous flow method (Skopp and McAllister, 1986):

$$\theta V\left(\frac{\partial C}{\partial t}\right) = AJ(C_{\mathrm{in}} - C_{\mathrm{eff}}) - V\rho_{\mathrm{B}}(\partial q/\partial t) \qquad (3.2)$$

where θ = volumetric water content, $m^3\ m^{-3}$

V = volume of filter holder, m^3

C = concentration of sorptive solution in filter holder, mol/liter
A = cross-sectional area of filter holder, m^2
J = flow rate, $m^3 \ min^{-1}$
C_{in} and C_{eff} = influent and effluent concentrations, respectively, $mol \ m^{-3}$
ρ_B = bulk density, $kg \ m^{-3}$
q = amount of reaction product expressed on either a mass or molar basis per unit mass of soil, $mol \ kg^{-1}$ or mol/liter kg^{-1}

The last term $(\partial q/\partial t)$ is an implicit reaction term expressing the unknown rate law. This term is written so that it is applicable to ion exchange, specific adsorption, precipitation or an enzyme-catalyzed reaction. This is possible since Eq. (3.2) represents a single equation in two unknowns (Skopp and McAllister, 1986).

The application of Eq. (3.2) assumes a thin disc which ignores any vertical concentration gradients. Thus, diffusion or hydrodynamic dispersion parallel to the average flow direction is not included.

It may be preferable to use the directly measurable quantity C rather than indirect ones such as q which are calculated from C, m, J, and t with an assumption of perfect mixing (Skopp and McAllister, 1986). If so, Eq. (3.2) must be solved. This necessitates an expression for dq/dt,

$$\frac{dq}{dt} = -k_{-1}q + k_1(r_c - q) \tag{3.3}$$

where r_c is the reaction capacity term ($mol \ kg^{-1}$) or (mol^{-1}). If one is studying cation exchange kinetics, then r_c represents the cation exchange capacity (CEC) in $mol \ kg^{-1}$. Equation (3.3) represents an apparent rate law and the k values determined are apparent. It should be pointed out that Eq. (3.3) is limited to situations where $q \ll r_c$. Skopp and McAllister (1986) present a number of solutions to Eq. (3.2) using Eq. (3.3) as well as other equations.

The analysis of data from a stirred-flow reactor is based on a mass balance equation similar to Eq. (3.2),

$$JC_{in} = JC_{eff} + V\frac{\partial C}{\partial t} + vV \tag{3.4}$$

where C = concentration of sorptive solution in the chamber, mol/liter;
V = volume of the chamber, m^3;
v = reaction rate, $mol \ m^{-3} \ s^{-1}$;

Equation (3.4) describes the mass balance relationship for one of the reactions being studied. It is also valid if the chamber is perfectly mixed or

$C = C_{\text{eff}}$. Subsequent equations will be expressed in terms of the measured value C.

In the original stirred-flow method (Denbigh, 1944), there were two or more openings for the flow of reactants and one opening for the flow of effluent. The effluent is a complex of reactants and products. With time, a steady state is established representing a balance between reactant additions in the influent and loss of reactant through reaction occurrence in the effluent. This steady state simplifies the mathematical treatment such that,

$$JC_{\text{in}} = JC - vV \tag{3.5}$$

Rearranging,

$$v = J(C_{\text{in}} - C)/V \tag{3.6}$$

When one studies kinetics of soil chemical processes, where solid surfaces are involved, the analysis of data using a stirred-flow reactor is different from that presented above. The main difference is the presence of one reactant, i.e., soil, clay mineral, or some other solid surface, whose mass is constant throughout the experiment. Thus, a steady state is established together with an equilibrium state when the net reaction rate is zero. Therefore, the analysis of data is not based on steady state conditions. However, continuous short-incremental measurements can be carried out, which enables analysis of non-steady state conditions.

Accordingly, the v term in Eq. (3.4) can be substituted by q, the quantity of ion or molecule sorbed or adsorbed in mol kg^{-1} to give

$$JC_{\text{in}} = JC + V\ \partial C/\partial t + m\ \partial q/\partial t \tag{3.7}$$

The analysis of the effluent data is based on testing a variety of rate laws by solving Eq. (3.7) repeatedly, each time using a different trial rate law. Then goodness-of-fit between the solution of Eq. (3.7) and the $C(t)$ data is used to select the appropriate rate law (Skopp and McCallister, 1986). Analytical solutions of Eq. (3.7) with a few rate laws (Langmuir model, first-order, and fractional-order rate laws) were presented by Skopp and McAllister (1986). The analytical solutions to the equations required linearity and the authors assumed that the maximum $q \gg$ the actual q in using the Langmuir and first-order equations. One coefficient was eliminated from the fractional-order rate laws. The assumptions of linearity and maximum $q \gg$ actual q are impractical in many soil chemical kinetics experiments, such as ion-exchange kinetics on soils and clay minerals. Fortunately, numerical solutions are available for systems described by nonlinear ordinary differential equations. In the author's laboratories the NAG software package was successfully employed to solve the Langmuir model and first-order equations free of the above-mentioned restrictions.

A practical problem in using the stirred-flow reactor involves choosing proper experimental conditions such that a set of $C(t)$ values are obtained that are significantly smaller than those obtained without a sorbent in the chamber (blank sample) and significantly different from those of an instantaneous reaction. With these concerns in mind, Seyfried *et al.* (1988) presented an empirical expression for the sorption or adsorption process which is given below

$$\frac{q}{EC} = D_g(C/C_T)[1 - \exp(-kt)] \tag{3.8}$$

where EC is the total exchange capacity in mol kg^{-1}, C_T is the total concentration of sorptive, or adsorptive, in mol m^{-3}, and D_g is a distribution coefficient determined from an exchange isotherm.

To solve Eq. (3.7) given Eq. (3.8), the latter can be derived with respect to time as follows, where the expression (D_g) $(EC)/C_T$ is replaced by the symbol Y,

$$\frac{\partial q}{qt} = Y[1 - \exp(-kt)]\frac{\partial C}{\partial t} + Yk[\exp(-kt)]C \tag{3.9}$$

Replacing $\partial q/\partial t$ on the right-hand side of Eq. (3.7) with the terms on the right-hand side of Eq. (3.9) and then rearranging gives the differential equation expression below,

$$\frac{\partial C}{\partial t} = \frac{J(C_{\text{in}} - C) - mkYC\exp(-kt)}{V - mY[1 - \exp(-kt)]} \tag{3.10}$$

Equation 3.10 can be solved numerically, using the NAG software package.

For known values of Y, a semiquantitative estimation of k is available by graphically fitting the data to theoretical curves where k values are varied as shown in Fig. 3.7.

The instantaneous condition is defined by Eq. (3.8) when the D_g value is so large that this equation is reduced to the following form (the expression $(D_g)(E_{EC})/C_T$ is replaced by the symbol Y),

$$q = YC \tag{3.11}$$

Equation (3.7) with equation 3.11 is solved analytically, under the initial condition; $t = 0$ $C = 0$, to give

$$C = C_i\{1 - \exp[-Jt/V + D_g m)] \tag{3.12}$$

Then, experiments can be run with and without adsorbent to calculate the relationship between q and t using Eq. (3.13)

$$q(t_i) = ([\Sigma(C_{\text{ns}_i} - C_{\text{s}_i})J\,\Delta t] + [c(t_i)_{\text{ns}} - c(t_i)_{\text{s}}]V)/m \tag{3.13}$$

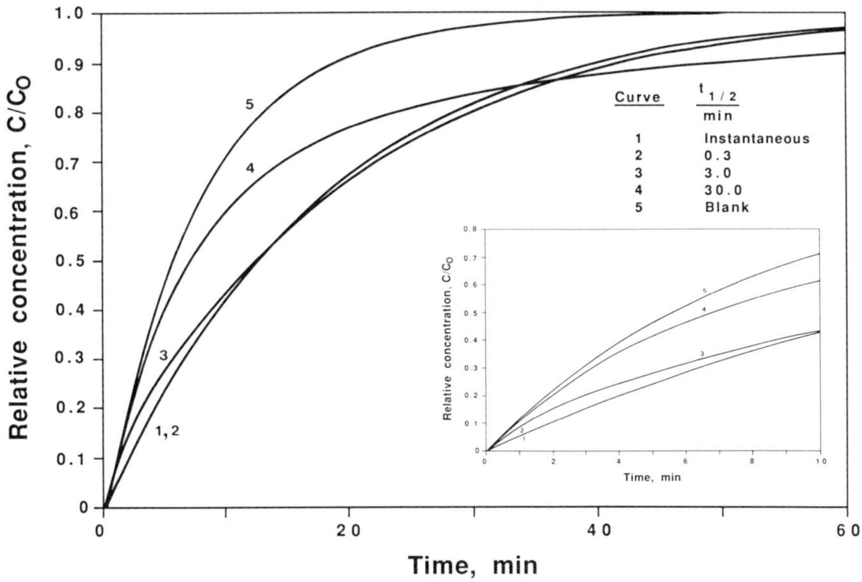

Figure 3.7. Theoretical curves to distinguish between blank, instantaneous, and different $t_{1/2}$ values using Eq. (3.2) and $J = 1.0$ ml min^{-1}, $V = 8.0$ ml, $m = 1.6$ g $CEC =$ 0.0625 cmol/kg, and $C_{in} = 10$ cmol/liter for (a) 10-min reaction time and (b) = 60-min reaction time.

where $q(t_i)$ is the amount of ion or molecule sorbed at time t, C is the concentration of ion or molecule in the collected solution, c is the concentration of sorptive in the reactor chamber, t_i is the time at the end of the sample collection period i, Δt is the length of the collection period, and subscripts ns and s denote no sorbed and sorbent in the reactor chamber, respectively.

COMPARISON OF KINETIC METHODS

Ogwada and Sparks (1986b) conducted a study investigating the effect of kinetic methodology and degree of agitation on rate parameters using five different methods. The methods included a static technique in which no mixing occurred, a continuous-flow method described earlier, a batch method in which the sample was agitated at 180 rpm on a reciprocating shaker, a stirred method where the mixture was stirred at 435 rpm, and a vortex batch technique whereby the mixture was rapidly mixed on a vortex mixer at 2240 rpm.

TABLE 3.3 Effect of Kinetic Method on Adsorption Rate Coefficients (k_a) in Systems Studied[a]

	k_a (min^{-1}) for kinetic method[b]				
Temperature (K)	Static	Continuous flow	Batch	Stirred	Vortex batch
Kaolinite					
283	0.030	0.033	1.304	1.321	1.444
298	0.034	0.037	2.383	2.838	3.482
313	0.037	0.045	3.764	3.784	4.981
Chester loam					
283	0.025	0.031	0.179	0.192	0.867
298	0.027	0.036	0.225	0.250	2.331
313	0.031	0.042	0.260	0.293	3.240
Vermiculite					
283	0.021	0.012	0.058	0.049	0.421
298	0.024	0.015	0.083	0.056	0.945
313	0.027	0.018	0.099	0.069	1.647

[a] From Ogwada and Sparks (1986b), with permission.

[b] The mixing rates were 180, 435, and 2240 rpm for the batch, stirred, and vortex batch methods, respectively.

Table 3.3 shows the effect of method on k_a values. The type of method clearly affected the k_a values and, although not shown, the time required for equilibrium in potassium adsorption to be reached. In earlier work, Ogwada and Sparks (1986a) had found that with the vortex batch method, diffusion was reduced significantly and the rate coefficients one obtained approximated reaction-controlled rate constants. The data in Table 3.3 show clearly that significant diffusion exists with the static and miscible displacement methods because of limited mixing.

As mixing increases—that is, with the batch and stirred techniques—the k_a values increase, and with kaolinite, are similar in magnitude to those obtained using vortex mixing. As noted earlier, the types of surfaces available for adsorption clearly affect the rapidity of the reaction. With kaolinite, only external surface sites are available and film diffusion (FD) should be rate-limiting. When the kaolinite mixture is perturbed, as in the batch or stirred systems, the film radius around the kaolinite particles is reduced and mass transfer across the film is rapid. Thus, the influence of FD in systems such as oxides and clay minerals like kaolinite that have only external surface sites can be diminished greatly with batch techniques where shaking or stirring is employed. In these systems, the rate coef-

TABLE 3.4 Effect of Kinetic Method on Energies of Activation for Adsorption (E_a) in Systems Studied[a]

	E_a (kJ mol^{-1}) for kinetic method[b]				
System	Static	Continuous flow	Batch	Stirred	Vortex batch
Kaolinite	5.05	7.57	26.06	26.01	30.59
Chester loam	5.25	7.46	9.19	19.92	32.58
Vermiculite	6.16	9.96	13.19	13.35	33.55

[a] From Ogwada and Sparks (1986b), with permission.
[b] The mixing rates were 180, 435, and 2240 rpm for the batch, stirred, and vortex batch methods, respectively.

ficients one obtains can be considered as approximating reaction-controlled rate constants.

On the other hand, with sorbents that contain internal sites for sorption that are not easily accessible (for example, vermiculitic interlayer sites), even mixing does not eliminate diffusion. This is illustrated in Table 3.3 for the Chester loam and vermiculite systems, both of which contain numerous interparticle sites because of their vermiculitic mineralogies. The k_a values for the batch and stirred methods were considerably lower than the vortex batch method, indicating that diffusion was still present. As can be seen in Table 3.4, diffusion was affected by mixing. The energy of activation values for adsorption (E_a) were strongly dependent on the kinetic technique employed. The E_a values were quite low for the static and miscible displacement techniques, indicating pronounced diffusion processes (Mortland and Ellis, 1959; Sparks, 1985, 1986). As mixing increased, the E_a values increased, indicating less mass transfer.

CONCLUSIONS

A number of methods can be used to study the kinetics of soil chemical processes. These include various types of batch and flow techniques. Each of these methods was described in this chapter and their advantages and disadvantages were discussed. It is obvious that none of them is a panacea for kinetic studies of heterogeneous systems such as soils. They each have strengths and weaknesses. It also appears that when most of these methods are used, apparent rate laws are being studied.

SUPPLEMENTARY READING

Barrow, N. J. (1983). A discussion of the methods for measuring the rate of reaction between soil and phosphate. *Fert. Res.* **40,** 51–59.

Carski, T. H., and Sparks, D. L. (1985). A modified miscible displacement technique for investigating adsorption-desorption kinetics in soils. *Soil Sci. Soc. Am. J.* **49,** 1114–1116.

Ogwada, R. A., and Sparks, D. L. (1986b). Kinetics of ion exchange on clay minerals and soil. I. Evaluation of methods. *Soil Sci. Soc. Am. J.* **50,** 1158–1162.

Phelan, P. J., and Mattigod, S. V. (1987). Kinetics of heterogeneously initiated precipitation of calcium phosphates. *Soil Sci. Soc. Am. J.* **51,** 336–341.

Skopp, J. (1986). Analysis of time dependent chemical processes in soils. *J. Environ. Qual.* **15,** 205–213.

Skopp, J., and Sparks, D. L. (1986). Chemical kinetics from a thin disc flow system: Theory. *Soil Sci. Soc. Am. J.* **50,** 617–623.

Sparks, D. L. (1985). Kinetics of ionic reactions in clay minerals and soils. *Adv. Agron.* **38,** 231–266.

Sparks, D. L. (1986). Kinetics of reactions in pure and in mixed systems. *In* "Soil Physical Chemistry" (D. L. Sparks, ed.), pp. 83–178. CRC Press, Boca Raton, Florida.

Sparks, D. L., Zelazny, L. W., and Martens, D. C. (1980). Kinetics of potassium desorption in soil using miscible displacement. *Soil Sci. Soc. Am. J.* **44,** 1205–1208.

Zasoski, R. G., and Burau, R. G. (1978). A technique for studying the kinetics of adsorption in suspensions. *Soil Sci. Soc. Am. J.* **42,** 372–374.

4

Kinetics and Mechanisms of Rapid Reactions on Soil Constituents Using Relaxation Methods

INTRODUCTION

The methods discussed in the previous chapters are primarily used to measure transport-controlled and diffusion-controlled reactions and to determine apparent rate laws. In this chapter, a discussion of rapid reactions with $t_{1/2}$ <10–20 s is presented. The determination of reaction rates for such systems is often tedious, since the reaction(s) is (are) often completed as soon as the sorptive and sorbent are mixed (Bernasconi, 1986). There are numerous reactions involving soil constituents and sorptives that are so rapid that the techniques discussed previously (Chapter 3) cannot be used to measure them. Reactions involving some metals and soil constituents are extremely rapid, and consequently can only be measured using rapid kinetic methods. Examples of kinetic techniques that can be used to measure rapid reactions between soil constituents and sorptives are pressure jump, stopped-flow, and electric field pulse. The theory behind each of these methods and their application to soil constituent reactions will be presented in this chapter.

The stopped-flow method has been extensively used to study biochemical reactions. With this technique, two reactants are rapidly mixed in a chamber, flow is then stopped rapidly a short distance away from the mixing chamber, and some physical property of the reaction system is measured with time. With this method, reactions with half-lives of a few milliseconds can be studied (Bernasconi, 1986).

However, the reaction rate is often more rapid than the mixing process when a flow method is used. In such a system, one must use a technique that does not depend on mixing. Relaxation kinetic methods (Table 4.1) would be appropriate for such a system (Eigen, 1954; Eigen and DeMaeyer, 1963). With these methods, the equilibrium of a reaction mixture is perturbed by a rapid alteration of some external factor such as pressure, temperature, or electric field strength. One then obtains kinetic information by measuring the "relaxation time" of the approach to a new equilibrium. The applied perturbation is usually small such that measurements are made close to the final equilibrium throughout the relaxation period. Consequently, concentration changes are small and one must utilize a very sensitive detection measurement like conductivity (Bernasconi, 1986). Because of the small perturbation, all rate expressions are reduced to first-order equations regardless of molecularity or reaction order. Thus, the rate equations are linearized, which greatly simplifies elucidation of complicated reaction mechanisms (Bernasconi, 1986).

Relaxation methods can be classified as either transient or stationary (Bernasconi, 1986). The former include pressure and temperature jump (p-jump and t-jump, respectively), and electric field pulse. With these methods, the equilibrium is perturbed and the relaxation time is monitored using some physical measurement such as conductivity. Examples of stationary relaxation methods are ultrasonic and certain electric field methods. Here, the reaction system is perturbed using a sound wave, which creates temperature and pressure changes or an oscillating electric field. Chemical relaxation can then be determined by analyzing absorbed energy (acous-

TABLE 4.1 Relaxation Methods Used in Kinetic Studies[a]

Method	Time range (s)	Method of detection
A. Transient methods		
1. Temperature jump	1–$10^{-6}(10^{-8})$	Spectrophotometric Fluorimetric Polarimetric
2. Pressure jump	5–10×10^{-5} (mechanical pressure release) 5×10^{-4}–5×10^{-7} (liquid shock wave)	Conductometric Spectrophotometric
3. Electrical field pulse	10^{-4}–10^{-8}	Conductometric Spectrophotometric
4. Concentration jump	10^{8}–10^{2} (conventional) 10^{3}–10^{-3} (stopped flow)	Spectrophotometric Fluorimetric and many others
B. Stationary methods		
5. Sound absorption and dispersion	10^{-5}–10^{-11} (overall time range for different acoustical techniques)	Power loss or frequency change: Resonance or reverberation (10^{4}–10^{6} Hz); light diffraction (10^{6}–10^{8} Hz); impulse echo (10^{6}–5×10^{8} Hz); Brillouin scattering
6. Dielectric dispersion	10^{-3}–10^{-12}	Power loss, capacitance change

[a] From Bernasconi (1976), with permission.

tical absorption or dielectric loss), or a phase lag that depends on the frequency of a forcing function (Bernasconi, 1986).

The t-jump method has been extensively used in chemistry and chemical engineering to study fast reactions. However, it has not been used to study soil constituent reactions. Bernasconi (1986) attributes its widespread use in other research fields to the facts that (1) almost every chemical equilibrium depends on temperature and can be perturbed with a temperature change; (2) several methods can be used to cause a t-jump, such as Joule heating, dielectric heating (microwave pulses), and optical heating (laser pulses); (3) t-jump measurements are amenable to several detection methods such as visible/ultraviolet (VIS/UV), optical rotation, fluorescence, Raleigh and Raman scattering, and conductivity; (4) reaction rates over a range of ~ 1 to 10^{-8} s can be determined; (5) t-jump apparatus are commercially available; and (6) one can combine t-jump and stopped-flow methods.

The p-jump technique has been applied to metal–soil constituent reactions and to ion exchange kinetics on zeolites. Specifics of these studies are presented in a later section. The p-jump method has several advantages over the t-jump technique (Knoche and Wiese, 1976). Solutions that contain biochemical compounds that can be obliterated with high electric fields and currents that are applied during Joule heating can be studied easily using the p-jump method. An additional advantage is that p-jump measurements can be repeated at faster intervals than with t-jump. In t-jump experiments, the solution temperature must return to its original value before another experiment can be carried out. This often takes about 5 min; however, p-jump experiments can be repeated every 0.5 min. A third advantage of the p-jump over the t-jump method is that longer relaxation times can be observed.

However, p-jump techniques are not without fault (Takahashi and Alberty, 1969). Most chemical reactions are less sensitive to pressure than to temperature alterations. Thus, a highly sensitive detection method such as conductivity must be employed to measure relaxation times if p-jump is used. Conductometric methods are sensitive on an absolute basis, but it is also fundamental that the solutions under study have adequate buffering and proper ionic strengths. In relaxation techniques, small molar volume changes result, and consequently, even if a low level of an inert electrolyte is present, conductivity changes may be undetectable if pressure perturbations of 5–10 MPa are utilized (Takahashi and Alberty, 1969).

Electric field pulse methods have only recently been used to study adsorption–desorption phenomena on heterogeneous systems such as soils and soil constituents. This ingenious application is discussed in a later section.

THEORY OF CHEMICAL RELAXATION

Linearization of Rate Equations

As mentioned earlier, one of the salient features of relaxation techniques for measuring fast reactions is the fact that due to small perturbations, all rate equations are reduced to first-order reactions. This linearization of rate equations is derived below and is taken entirely from Bernasconi (1976).

Consider a chemical equilibrium such as

$$\mathrm{A} + \mathrm{B} \underset{k_{-1}}{\overset{k_1}{\rightleftharpoons}} \mathrm{Y} \tag{4.1}$$

The equilibrium state is perturbed by adding more product, or by dilution, pH changes, or alterations in temperature or pressure, and the system adjusts itself to the altered set of external factors.

This adjustment process results in a change in the concentrations of some or all of the species. The rate of the adjustment to new equilibrium conditions or the rate of "chemical relaxation" is determined by the rate of the reactions that make up the equilibrium. By measuring the relaxation rate, one can obtain information that can be used to determine k_1 and k_{-1}.

Assume that the general rate equation

$$-dc_A/dt = -dc_B/dt = dc_Y/dt = k_1 c_A c_B - k_{-1} c_Y \tag{4.2}$$

which explains rate behavior of the system under any set of conditions and where c_A, c_B, and c_Y refer to the concentrations of reactants A and B and product Y. At equilibrium Eq. (4.2) becomes

$$-dc_A/dt = (k_1 \bar{c}_A)(\bar{c}_B - k_{-1}\bar{c}_Y) = 0 \tag{4.3}$$

The bar over the concentration symbols indicates equilibrium concentrations given by the mass law expression

$$\bar{c}_Y/(\bar{c}_A \bar{c}_B) = k_1/k_{-1} = \mathrm{K}_{eq} \tag{4.4}$$

In a relaxation experiment the system relaxes from an initial equilibrium state to a final equilibrium state, and one can propose the following symbols:

$\bar{c}^i_A$, $\bar{c}^i_B$, $\bar{c}^i_Y$: Equilibrium concentration at initial pressure P_i

k^i_1, k^i_{-1}; $K^i_{eq} = k^i_1/k^i_{-1}$: Rate and equilibrium constants at p_i

$\bar{c}^f_A$, $\bar{c}^f_B$, $\bar{c}^f_Y$: Equilibrium concentrations at final pressure $p_f = p_i + \Delta p$

k^f_1, k^f_{-1}; $K^f_{eq} = k^f_1/k^f_{-1}$: Rate and equilibrium constants at $p_f = p_i + \Delta p$

In terms of these symbols Eq. (4.3) can be written as

$$-dc_A/dt = (k^i_1 \bar{c}^i_A)\bar{c}^i_B - k^i_{-1}\bar{c}^i_Y = 0 \tag{4.5}$$

or

$$dc_A/dt = (k^f_1 \bar{c}^f_A)\bar{c}^f_B - k^f_{-1}\bar{c}^f_Y = 0 \tag{4.6}$$

Equation (4.5) describes the reaction state before the pressure jump, while Eq. (4.6) explains the reaction after chemical relaxation.

One assumes that the pressure jump is instantaneous and that it is on a time scale similar to the actual chemical relaxation. The concentrations immediately after the pressure jump or before relaxation commences are those referring to the initial equilibrium state (before the pressure jump). However, one is at p_f and the rate constants are k^f_1 and k^f_{-1}. The rate equation then becomes (Bernasconi, 1976)

$$-dc_A/dt = (k^f_1 \bar{c}^i_A)\bar{c}^i_B - k^f_{-1}\bar{c}^i_Y \tag{4.7}$$

For $K^f_{eq} \neq K^i_{eq}$, which is usually the case, one has $dc_A/dt \neq 0$ in Eq. (4.7).

After a period of time, relaxation occurs and the concentrations of A, B, and Y approach their final equilibrium values $\bar{c}_A^f$, $\bar{c}_B^f$, and $\bar{c}_Y^f$. Equation (4.7) is then a special case of the more general equation

$$-dc_A/dt = k_1^f c_A c_B - k_{-1}^f c_Y \tag{4.8}$$

which may refer to a state where relaxation has occurred but is not complete.

Formally, one expresses the time-dependent concentrations c_A, c_B, and c_Y in terms of their equilibrium values and their time-dependent deviations from these equilibrium values:

$$c_A = \bar{c}_A^f + \Delta c_A \tag{4.9a}$$

$$c_B = \bar{c}_B^f + \Delta c_B \tag{4.9b}$$

$$c_Y = \bar{c}_Y^f + \Delta c_Y \tag{4.9c}$$

According to the principle of mass balance,

$$c_A + c_Y = \bar{c}_A^f + \bar{c}_Y^f \tag{4.10a}$$

$$c_B + c_Y = \bar{c}_B^f + \bar{c}_Y^f \tag{4.10b}$$

Substituting Eqs. (4.9) for c_A, c_B, and c_Y, respectively, yields

$$\Delta c_A + \Delta c_Y = 0 \tag{4.11a}$$

$$\Delta c_B + \Delta c_Y = 0 \tag{4.11b}$$

or

$$\Delta c_A = \Delta c_B = -\Delta c_Y = x \tag{4.11c}$$

Thus, Eqs. (4.9) become

$$c_A = \bar{c}_A^f + x \tag{4.12a}$$

$$c_B = \bar{c}_B^f + x \tag{4.12b}$$

$$c_Y = \bar{c}_Y^f - x \tag{4.12c}$$

Since the relation

$$\begin{aligned} dc_A/dt = dc_B/dt = -dc_Y/dt &= d(\bar{c}_A^f + x)/dt \\ &= d\bar{c}_A/dt + dx/dt = dx/dt \end{aligned} \tag{4.13}$$

holds (note $d\bar{c}_A/dt = 0$), Eq. (4.8) can be written as

$$\begin{aligned} dx/dt &= -k_1^f(\bar{c}_A^f + x)(\bar{c}_B^f + x) + k_{-1}^f(\bar{c}_Y - x) \\ &= -k_1^f \bar{c}_A^f \bar{c}_B^f + k_{-1}^f \bar{c}_Y^f - k_1^f (x)^2 \\ &\quad -[k_1^f(\bar{c}_A^f + \bar{c}_B^f) + k_{-1}^f]x \end{aligned} \tag{4.14}$$

The first two terms disappear because of Eq. (4.6) and one obtains

$$dx/dt = -[k_1^{\mathrm{f}}(\bar{c}_{\mathrm{A}}^{\mathrm{f}} + \bar{c}_{\mathrm{B}}^{\mathrm{f}}) + k_{-1}^{\mathrm{f}}]x - k_1^{\mathrm{f}}(x)^2 \tag{4.15}$$

which can be simplified if only small equilibrium perturbations are considered. A perturbation is small when $|x| = |\Delta c_j| \ll \bar{c}_j^{\mathrm{f}}$ immediately after the perturbation for all species taking place in the equilibrium. Thus, the square term $k_1^{\mathrm{f}}(x)^2$ is very small compared to the other terms and Eq. (4.15) becomes

$$dx/dt = -[k_1(\bar{c}_{\mathrm{A}} + \bar{c}_{\mathrm{B}}) + k_{-1}]x \tag{4.16}$$

The deletion of the square term is called the *linearization of the rate equation*. The superscript f from Eq. (4.16) is deleted to simplify matters.

Equation (4.16) can now be written as

$$dx/dt = -(1/\tau)x \tag{4.17}$$

with

$$1/\tau = k_1(\bar{c}_{\mathrm{A}} + \bar{c}_{\mathrm{B}}) + k_{-1} \tag{4.18}$$

Equations (4.16) and (4.17) are independent of the sign of x. The original equilibrium could be shifted to the right or the left; the rate of approach to the new equilibrium is identical. One could have thus chosen $x = \Delta c_{\mathrm{Y}}$ rather than $x = \Delta c_{\mathrm{A}}$. Thus, rather than Eq. (4.17), one can write

$$d(\Delta c_j)/dt = -(1/\tau)\,\Delta c_j \tag{4.19}$$

where j refers to any of the reacting species A, B, or Y. This is a salient feature of chemical relaxation only if the perturbation is small. For large perturbations dx/dt *does* depend on the sign of x; see Eq. (4.15). This is because a change in sign of x changes only the sign of the dx/dt and of the $[k_1(\bar{c}_{\mathrm{A}} + \bar{c}_{\mathrm{B}}) + k_{-1}]x$ terms.

Relaxation Time

Assuming that the system $\mathrm{A} + \mathrm{B} \rightleftharpoons \mathrm{Y}$ is functional, one can linearize the rate equation of any one-step equilibrium. Equations (4.17) and (4.19) are first-order equations that result from linearizing the rate equation of a one-step equilibrium like Eq. (4.1). The symbol τ is the relaxation time of the system.

Integration of Eq. (4.17)

$$\int_{x_0}^{x} x^{-1}\,dx = \tau^{-1}\int_0^t dt$$

leads to

$$\ln(x/x_0) = -t/\tau \tag{4.20}$$

or

$$x = x_0 \exp(-t/\tau) \tag{4.21}$$

with

$$x_0 = \bar{c}_A^i - \bar{c}_A^f \tag{4.22}$$

From Eq. (4.19) one obtains

$$\Delta c_j = \Delta c_j^0 \exp(-t/\tau) \tag{4.23}$$

or

$$c_j = \bar{c}_j^f + \Delta c_j^0 \exp(-t/\tau) \tag{4.24}$$

with

$$\Delta c_j^0 = \bar{c}_j^i - \bar{c}_j^f \tag{4.25}$$

Figure 4.1 shows the relationship between the above quantities. The relaxation time corresponds to the time needed for Δc_j to decrease by a factor (base of natural logarithims, $e = 2.718$). This can be accomplished by setting $t = \tau$ in Eq. (4.23), which gives

$$\Delta c_j = \Delta c_j^0/e = \Delta c_j^0/2.718 = 0.368\ \Delta c_j^0$$

Some investigators have expressed x or Δc_j with reference to the equilib-

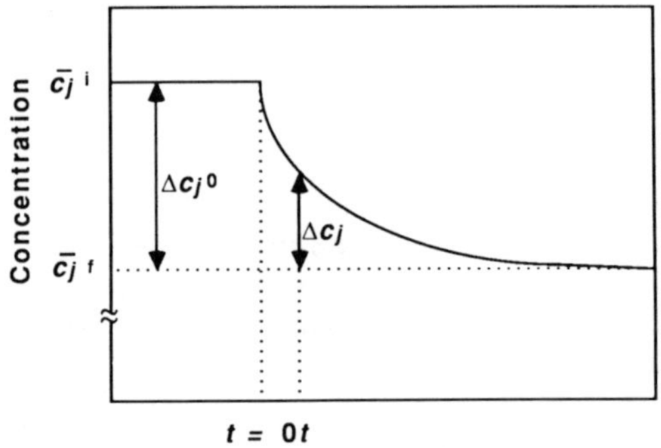

Figure 4.1. Chemical relaxation according to Eq. (4.24). [From Bernasconi (1976), with permission.]

rium condition before perturbation. One then obtains

$$c_A = \bar{c}^i_A + \Delta c_A = \bar{c}^i_A + x = \bar{c}^f_A + x - x_\infty \tag{4.26a}$$

$$c_B = \bar{c}^i_B + \Delta c_B = \bar{c}^i_B + x = \bar{c}^f_B + x - x_\infty \tag{4.26b}$$

$$c_Y = \bar{c}^i_Y + \Delta c_Y = \bar{c}^i_Y - x = \bar{c}^f_Y - x + x_\infty \tag{4.26c}$$

with

$$x = c_A - \bar{c}^i_A \tag{4.27}$$

$$x_\infty = \bar{c}^f_A - \bar{c}^i_A \tag{4.28}$$

Linearization of the rate equation gives

$$dx/dt = -(1/\tau)(x - x_\infty) = (1/\tau)(x_\infty - x) \tag{4.29}$$

while integration

$$\int_0^t \frac{dx}{x_\infty - x} = \frac{1}{\tau}\int_0^t dt$$

leads to

$$\ln \frac{x_\infty - x}{x_\infty} = \frac{-t}{\tau} \tag{4.30}$$

or

$$x = x_\infty\,[1 - \exp(-t/\tau)] \tag{4.31}$$

The reciprocal relaxation time τ^{-1} is analogous to the rate constant calculated according to first-order kinetics. Usually τ is determined using Eq. (4.20), or $\log_{10}$ and rearranging,

$$\log x = \log x_0 - t/(2.303\tau) \tag{4.32}$$

Thus, a plot of log x versus time should be linear with a slope of $-1/2.303\tau$.

The determination of τ is made easier because of several aspects of Eq. (4.32). It is not necessary to know the real zero point ($t = 0$) of the relaxation curve; moreover, a physical property such as absorbance or conductance related to concentration is measured in relaxation studies rather than actual concentrations. With first-order reactions, one does not need to know the proportionality constant between the physical property and the concentration of the respective species. This is valid only if one or all of the species present in the system contributes to the physical property (Bernasconi, 1976). Another basic feature of chemical relaxation that should be mentioned is that the forward and backward reactions of Eq. (4.1) contribute additively to τ^{-1} (Bernasconi, 1976). Thus, it is the faster of the two processes which contributes most to τ^{-1}.

Evaluation and Determination of Relaxation Times and Rate Constants in Single-Step Systems

The reciprocal relaxation time (τ^{-1}) and the forward and backward rate constants (k_1 and k_{-1}, respectively) can be found using Eq. (4.18):

$$\tau^{-1} = \frac{1}{\tau} = k_1(\bar{c}_A + \bar{c}_B) + k_{-1}$$

From a linear plot of $1/\tau$ versus $(\bar{c}_A + \bar{c}_B)$ one can calculate k_1 and k_{-1} from the slope and intercept, respectively (Fig. 4.2).

To get very accurate values for k_{-1} and k_1, the experiment should be carried out over a wide concentration range. The highest degree of accuracy for k_{-1} is found from data where $k_{-1} > (\gg)\ k_1(\bar{c}_A + \bar{c}_B)$. Good precision in k_1 results when concentrations are extended such that $k_1(\bar{c}_A + \bar{c}_B) > (\gg)\ k_{-1}$ (Bernasconi, 1976).

The stoichiometric rather than the equilibrium concentrations of A and B are often known, unless one can measure c_A and c_B directly or the equilibrium constant for the reaction is known independently (Bernasconi, 1976). Thus, Eq. (4.18) may have little utility. However, there are several ways to increase its usefulness (Bernasconi, 1976). One can maintain the concentration of one of the reactants quasi-constant by having it in large excess over the other reactant (pseudo-first-order conditions).

Therefore, for $[A]_0 \gg [B]_0$, where $[A]_0 = \bar{c}_A + \bar{c}_Y$ and $[B]_0 = \bar{c}_B + \bar{c}_Y$, $\bar{c}_A \simeq [A]_0 \gg \bar{c}_B$ and Eq. (4.18) becomes

$$\tau^{-1} = k_1[A]_0 + k_{-1} \tag{4.33}$$

Thus, k_1 and k_{-1} are found from a plot of τ^{-1} versus $[A]_0$. By keeping $[A]_0$

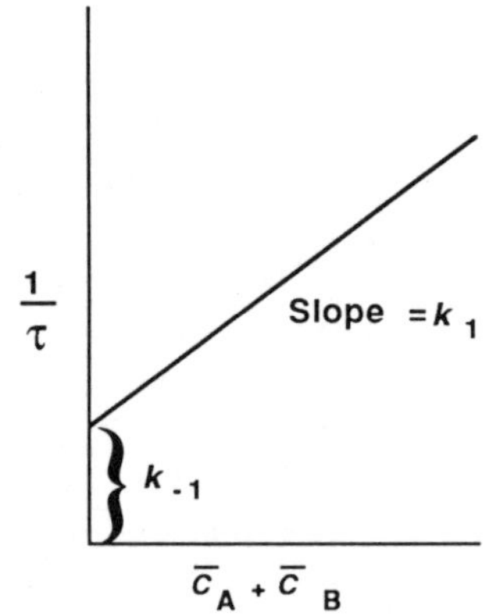

Figure 4.2. Determination of rate constants from the concentration dependence of $1/\tau$ for the $A + B \rightleftharpoons Y$ system based on Eq. (4.18). [From Bernasconi (1976), with permission.]

quasi-constant, one is not restricted to only small equilibrium perturbations (Bernasconi, 1976). The fact that perturbations of any degree are allowed under these conditions can be useful experimentally. This is an important advantage in using relaxation techniques, since large perturbations result in larger concentration changes, which improve overall accuracy. This results because there is no square term in the rate equation, which can be shown if one defines a pseudo-first-order rate coefficient as $k_x = k_1[A]_0$. Then, Eq. (4.14) can be given as

$$dx/dt = -k_x(\bar{c}_B + x) + k_{-1}(\bar{c}_Y - x) = -(k_x + k_{-1})x \quad (4.34)$$

which has no square term.

It should be pointed out that choosing one of the reactants in large excess may not always be desirable. As an example, if $k_1 \geq 10^4 k_{-1}$, then to get an accurate k_{-1} value, the concentration of the excess reactant (A) would have to be 10^{-4} *M* or lower. Then, the concentration of B should be 10^{-5} *M* or lower. The latter concentration may be too low to measure concentration changes with relaxation.

On the other hand, the absolute values of k_1 could be high, and consequently, the term $k_1[A]_0$ would be so large with pseudo-first-order kinetic conditions that τ would be extremely small. One could lower the concentration of A to that of B so that τ is measurable (Bernasconi, 1976). In such situations the best thing to do is to choose $[A]_0 = [B]_0$ and Eq. (4.18) would become

$$(\tau^{-1})^2 = 4k_1k_{-1}[A]_0 + (k_{-1})^2 \quad (4.35)$$

Plotting $(\tau^{-1})^2$ versus $[A]_0$ would give a straight line with a slope of $4k_1k_{-1}$ and an intercept of $(k_{-1})^2$.

If neither of the above situations is possible, as for pseudo-first-order conditions or $[A]_0 = [B]_0$, k_1 and k_{-1} can be determined using an iteration method (Bernasconi, 1976).

PRESSURE-JUMP (p-JUMP) RELAXATION

Historical Perspective

Pressure-jump relaxation methods (Takahashi and Alberty, 1962; Eigen and DeMaeyer, 1963; Hoffman *et al.*, 1966; Knoche, 1974; Gruenewald and Knoche, 1979; Yasunaga and Ikeda, 1986) and theory (Takahashi and Alberty, 1969; Bernasconi, 1976) have been reviewed extensively, and the reader is referred to these references for in-depth discussions. The p-jump methods are based on the fact that chemical equilibria are dependent on pressure, as is shown below (Bernasconi, 1976):

$$\left(\frac{\partial \ln K_{eq}}{\partial p}\right)_T = -\frac{\Delta V}{RT} \tag{4.36}$$

where ΔV is the standard molar volume change of the reaction (liters), p is pressure (MPa), and R is the molar gas constant. One can now write

$$\frac{\Delta K_{eq}}{K_{eq}} = \frac{-\Delta V}{RT}\Delta p \tag{4.37}$$

A sudden pressure release or application of pressure can be employed to cause the pressure jump. Ljunggren and Lamm (1958) described the first pressure-jump apparatus, which consisted of a sample cell connected to a nitrogen tank. With this apparatus, a pressure increase to 15.2 MPa could be obtained in ~50 ms by quickly opening the valve. Chemical relaxation was monitored conductometrically.

In 1959 Strehlow and Becker developed a pressure-jump apparatus that enclosed a conductivity cell containing the reaction solution, and a reference cell under xylene in an autoclave. The reaction and reference solutions were pressurized to about 6.1 MPa with compressed air. By the blow of a steel needle, a thin metal disk used to close the autoclave was punctured and the pressure was released within about 60 s.

A number of modifications of the Strehlow and Becker (1959) original p-jump apparatus have been described and employed (Strehlow and Becker, 1959; Hoffmann *et al.*, 1966; Takahashi and Alberty, 1969; Macri and Petrucci, 1970; Knoche, 1974; Knoche and Wiese, 1974; Davis and Gutfreund, 1976). Patel *et al.* (1974) described a p-jump cell constructed entirely from Plexiglass; which improved thermostatting so that the thermostatting fluid is in direct contact with the cells and the sample solution is clearly visible at all times. Smaller relaxation times than those using the devices cited above have been measured by causing pressure changes with shock waves (Jost, 1966), standing sound waves (Wendt, 1966), and electromechanical techniques (Hoffman and Pauli, 1966).

Pressure-Jump Apparatus

An adaptation of the p-jump device described by Strehlow and Becker (1959) was introduced by Knoche and Wiese (1974) and a description of it is given below. This apparatus has been used by numerous investigators to study fast reactions on soil constituents, and a modification of it is commercially available.

Relaxation times of $\geq$30 μs can be measured using the p-jump unit

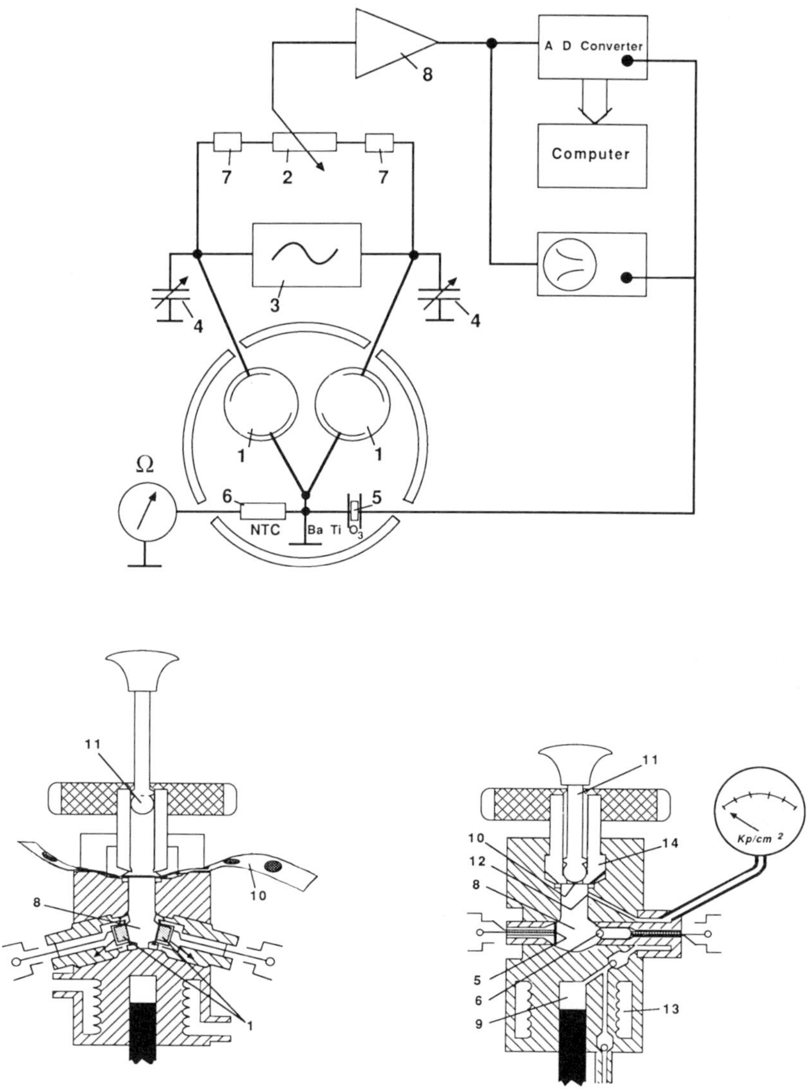

Figure 4.3. Schematic diagram and sectional views of the autoclave of the pressure-jump apparatus of Knoche and Wiese (1974): 1, conductivity cells; 2, potentiometer; 3, 40-kHz generator for Wheatstone bridge; 4, tunable capacitors; 5, piezoelectric capacitor; 6, thermistor; 7, 10-turn helipot for tuning bridge; 8, experimental chamber; 9, pressure pump; 10, rupture diaphragm; 11, vacuum pump; 12, pressure inlet; 13, heat exchanger; 14, bayonet socket. [From Knoche and Wiese (*1974*), with permission.]

(Fig. 4.3) introduced by Knoche and Wiese (1974). Particular care was taken in the development of this apparatus to reduce mechanical disturbances after the p-jump, which frequently cause large dead times (time that elapses from maximum applied pressure to pressure of 0.1 MPa). Referring to Fig. 4.3, two identical conductivity cells (1) are mounted inside the autoclave where rapid pressure changes can occur. One of the cells is filled with the solution under study while the second one contains a solution with the same electrical conductivity but exhibiting no relaxation during the time scale under study. By comparing the resistance change of the two cells, disturbances created by temperature equilibration cancel. The conductivity cells and the potentiometer (2) create a Wheatstone bridge. The voltage source is a 40-kHz generator (3). Two tunable capacitors (4) are connected in parallel to the cells, which enables capacitive tuning. The bridge signal is amplified, then displayed on the oscilloscope screen and digitized. The digitized signal is input for a small computer that evaluates the relaxation times.

The autoclave is closed with a burst diaphragm, which ruptures spontaneously at a pressure of about 13.1 MPa. Before the measurement, the bridge is tuned at a pressure of 0.1 MPa. The pressure is then increased slowly until the burst diaphragm blows out. The pressure decreases within about 100 μs to ambient, and the voltage peak of the piezoelectric capacitor (5) triggers the oscilloscope and the digitizer. The trace on the oscilloscope screen now shows how the investigated solution regains equilibrium at the ambient pressure of 0.1 MPa. Simultaneously, the signal is digitized, and the computer calculates the relaxation time and the amplitude of the relaxation effect. The electric resistance of the NTC resistor (6) is measured by a bridge circuit, which yields the temperature in the autoclave. Using a 10-turn helipot (7), the bridge is tuned to zero signal, and the temperature can be measured to an accuracy of $\pm$273.1 K. The calibration curve of the NTC resistor only has to be determined once. Details of the electrical circuit, the digitizing of the signal, and the data processing can be found in other references (Knoche and Wiese, 1974, Refs. 3–5).

Knoche and Wiese (1974) made a number of alterations to the autoclave (Fig. 4.3) originally proposed by Strehlow and Becker (1959). The energy released at the pressure jump is partly needed to break the rupture disk but can cause the autoclave to oscillate. This disturbs the determination of the cell resistances. To minimize the energy, the experimental chamber (8) volume was reduced and the pressure pump (9) was built as an integrated part of the autoclave to reduce all supply lines to a minimum. With this autoclave, water was used as the pressure transducing liquid instead of than kerosene (Strehlow and Becker, 1959), to reduce the compressibility. The conductivity cells were also mounted on a small incline so that no air bubbles are trapped in the experimental chamber, which can cause pressure oscillations after the pressure jump. Other improvements in the autoclave of Knoche and Wiese (1974) included avoiding acoustical noise disturbances during the pressure jump by reducing pressure using a pump (11), making the pressure inlet (12) close beneath the rupture disk so that few air bubbles would reach the experimental chamber, and regulating the autoclave temperature by thermostatting the liquid streaming rapidly through the heat exchanger (13).

Conductivity and Optical Detection Using p-Jump Relaxation

Conductivity Detection. Pressure-jump measurements can be detected using either optical or conductivity detection. However, conductivity detection is usually preferred since the equilibrium displacement following p-jump is usually small (Bernasconi, 1976). Conductometric detection has been exclusively used by researchers investigating the rapid kinetics of reactions on soil constituents (to be discussed later) because of the high sensitivity, obtained using conductivity and because suspensions are studied. Optical detection would not be desirable for suspensions.

Conductivity detection is excellent for ionic or dipolar systems. It is optimal only if the ions under study are contained in solution or suspension, since the sensitivity of detection decreases when other electrolytes or acid–base buffers are added (Bernasconi, 1976). However, Strehlow and Wendt (1963) obtained good precision even in systems where 98% of the conductance was ascribable to inert electrolytes. One way to minimize extra conductance is to add salts with ions of low mobility such as tetraalkylammonium ions, rather than sodium or potassium ions.

The specific conductance σ of an electrolyte solution is given as (Bernasconi, 1976)

$$\sigma = \frac{F}{1000}\sum c_j|z_j|u_j = \frac{F}{1000}\rho\sum m_j|z_j|u_j \tag{4.38}$$

where F is the Faraday constant, z_j is the valence of ion j, c_j is the molar and m_j the molal concentration of ion j, u_j is its electrical mobility (cm^2 V^{-1} s^{-1}), and ρ is the density of the solution.

It is better to express concentration in molality to separate the concentration changes effected by chemical relaxation from those created by volume or density changes. For small perturbations one can write (Eigen and DeMaeyer, 1963)

$$\Delta\sigma = \frac{F}{1000}(\rho\sum|z_j|u_j\Delta m_j + \rho\sum|z_j|m_j\Delta u_j + \sum|z_j|m_j(u_j\Delta\rho) \tag{4.39}$$

The first term on the right-hand side of Eq. (4.39) indicates chemical relaxation, while the remaining terms are physical effects, such as changes of ionic mobility and density due to pressure and temperature changes. The temperature change can be eliminated by using a reference cell filled with a nonrelaxing solution with the same temperature dependence of the conductivity as the sample cell (Knoche and Wiese, 1974).

Optical Detection. Optical detection can be used to assess concentration changes and can be used for a number of different systems. One of its advantages is that one is not restricted to studying ionic reactions (Knoche and Wiese, 1976). Changes in optical properties can be followed very rapidly and with excellent sensitivity using photoelectric transducers or photomultipliers (Eigen and DeMaeyer, 1963). Absorption spectrometry can be used with temperature as the forcing variable (Czerlinski, 1960). Fluorescence spectrometry may be used to increase sensitivity at low concentrations (Czerlinski, 1960), or refractometric or polarometric techniques or measurement of optical rotation may be used (Eigen and DeMaeyer, 1963).

Knoche and Wiese (1976) described a p-jump apparatus using an optical absorption detection system, but relaxation times $<3 \times 10^{-4}$s could not be observed and the detection was not as sensitive as when electrical conductivity was measured. A p-jump apparatus based on Strehlow's design (Strehlow and Becker, 1959) but using spectrophotometric detection was also described by Goldsack *et al.* (1969).

Other Methods of Detection. Other ways to detect p-jump relaxations include thermal properties of the reaction system using a rapid calorimetric method in which the heat of reaction is measured. The main problem with this method is that the time resolution is not very high.

Evaluation of p-Jump Measurements

Pressure-jump measurements can be evaluated using nondigital or digital techniques. Since nondigital methods are time-consuming and arduous, digital techniques are greatly preferred. A brief discussion of both techniques is given below.

Nondigital Techniques. The older method of evaluating p-jump relaxation experiments was to photograph the trace on an oscilloscope screen and then to plot the measured amplitudes against time on semilogarithmic paper. A typical oscilloscope picture looks like a ringing pattern whose

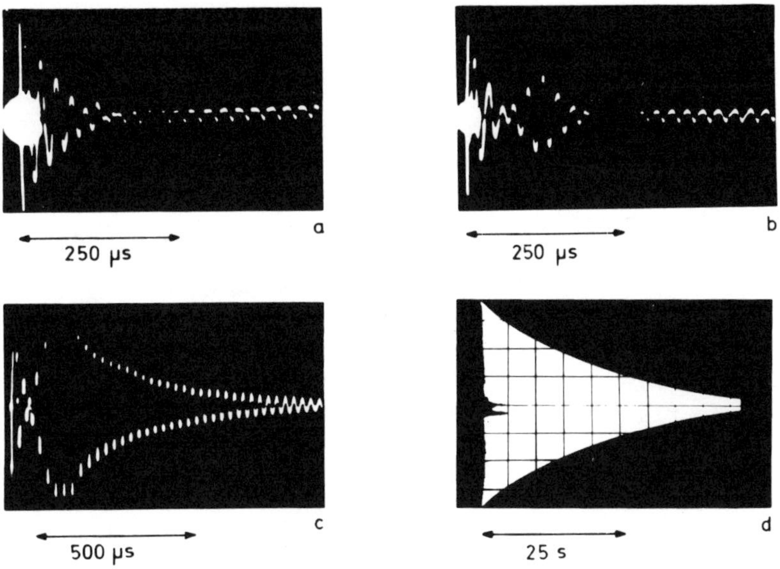

Figure 4.4. Typical oscillograms of pressure-jump experiments. Relative change in conductivity for pressure-jumps of 13.1 MPa in solutions of 0.05 *M* $InCl_3$, pH = 3.25. (a) At 383 K showing only pressure decay. (b) At 300.5 K, $\tau = 50 \pm 15$ μs. (c) At 273.7 K, $\tau = 215 \pm 10$ μs. (d) Solution of 0.10 *M* α-ketoglutaric acid, pH 1.69, at 274 K, $\tau = 25.8$ s. [From Knoche and Wiese (1974), with permission.]

frequency is the oscillating frequency of the bridge and whose envelopes correspond to the exponential relaxation decay (Fig. 4.4). Crooks *et al.* (1970) improved this method by eliminating the amplitude measurements and the error introduced by the imperfect oscilloscope. Crooks *et al.* (1970) photographed the relaxation trace and projected it onto the same oscilloscope screen that had been used for the experiments. An exponentially decaying time function is then generated on the scope with a suitable RC network. The amplitude and time constant of this function are changed until the projected curve on the scope is matched. This method has also been extended for two relaxation times. It is better than the old method, but darkroom work is still necessary. Also, the results are not available until the experiment is completed (Knoche and Strehlow, 1979).

Digital Techniques. Digital techniques greatly facilitate p-jump relaxation data analyses and are currently available commercially through Dia-Log (distributed by Interactive Radiation, Inc. Northvale, New Jersey). A more complete discussion of commercially available p-jump units and digitizers is given later.

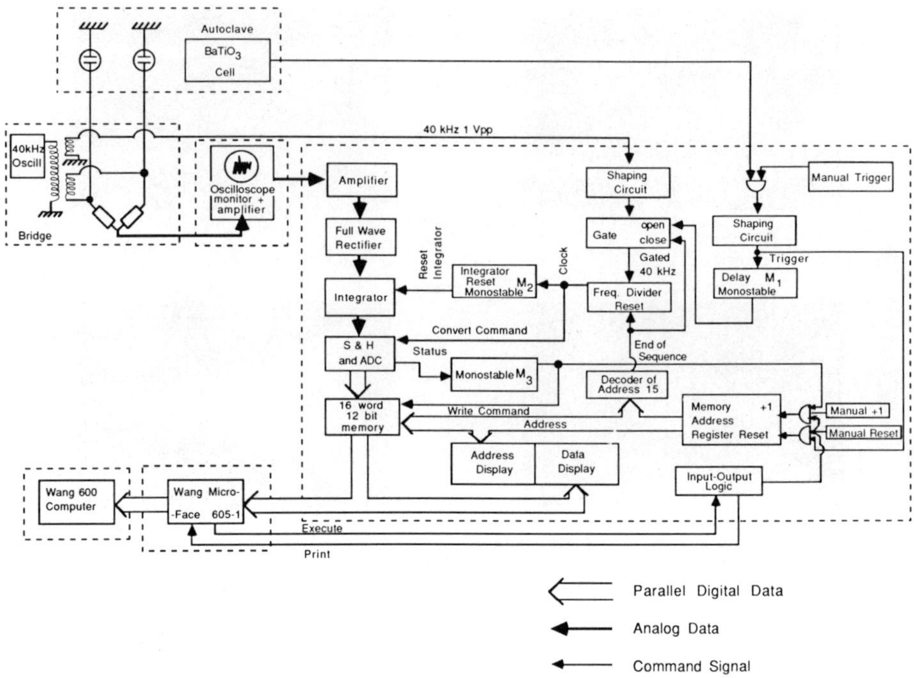

Figure 4.5. Block diagram of pressure-jump relaxation apparatus with digitizing interface. [From Krizan and Strehlow (1974), with permission.]

Krizan and Strehlow (1974) described a relatively simple digitizing interface for use in p-jump relaxation experiments which improves the accuracy and rapidity of data acquisition. A block diagram of the p-jump apparatus with digitizing interface is shown in Fig. 4.5. With this interface, one can sample by repeating the experiments and adding data to a small on-line computer.

Knoche and Strehlow (1979) describe in some detail data capture and processing in chemical relaxation measurements. A BASIC computer program and programs for the microcomputers Wang 600, Wang 720, and Hewlett-Packard (HP) 9815 and for the pocket calculator HP 67/97 (computing time ~5 min) are available. The hardware can be obtained from Dia-Log Co, Harffstra, βe34, D-4000 Düsseldorf 13, West Germany.

Commercially Available p-Jump Units

As of this writing, the availability of commerical p-jump units is limited to units produced by Dia-Log Co. and distributed by Inrad Interactive

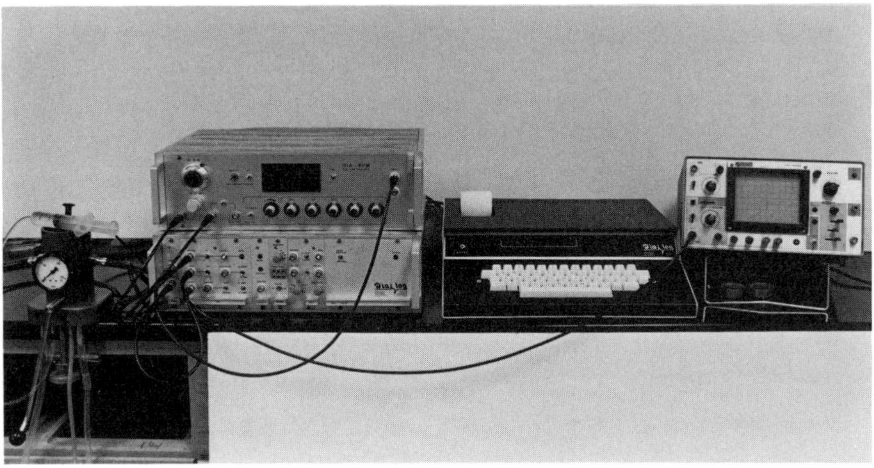

Figure 4.6. Dia-Log Co. pressure-jump unit with conductivity detection. (Used with permission of the manufacturer.)

Radiation, Inc. (181 Legrand Ave., Northvale, New Jersey 07647), and by Hi-Tech Scientific Limited (Brunel Road, Salisbury, Wiltshire, England SP27PU). A short discussion of these two units is given below.

Dia-Log Co. manufactures p-jump units with optical conductivity detection capabilities. A photograph of the p-jump unit with conductivity detection is shown in Fig. 4.6. Relaxation times of 50 μs–100 s can be measured. The conductivity range is 200 S $m-^{1}$ to 0.05 S m^{-1}, the temperature range is 273–343 K, and a sample volume of 0.5 ml or more can be used with a readout digitizer that has a memory of up to 256 values. It provides automatic data processing and data reduction with a microprocessor. The data can also be evaluated with PET, HP 67, or Wang 600 and 720 hand-held calculators.

The p-jump unit produced by Hi-Tech Limited (PJ-55 pressure-jump) is based on a design by Davis and Gutfreund (1976) and is shown in Fig. 4.7, with a schematic representation in Fig. 4.8. A mechanical pressure release valve permits observation after 100 μs. There is no upper limit to observation time. Changes in turbidity, light absorption, and fluorescence emission can be measured in the range of 200–850 nm. The PJ-55 is thermostated by circulating water from an external circulator through the base of the module. The temperature in the cell is continuously monitored with a thermocouple probe. A hydraulic pump assembly is used to build up a pressure of up to 40.4 MPa. A mechanical valve release causes the pressure build-up to be applied to the solution in the observation cell. The instrument has a dead time of 100 μs. A fast response UV/fluorescence

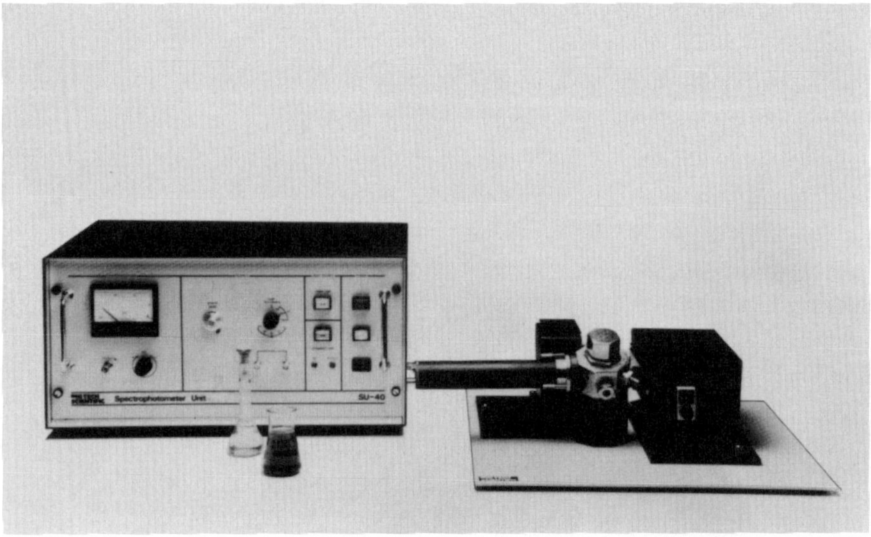

Figure 4.7. Hi-Tech Limited pressure-jump (PJ-55) unit. (Used with permission of the manufacturer.)

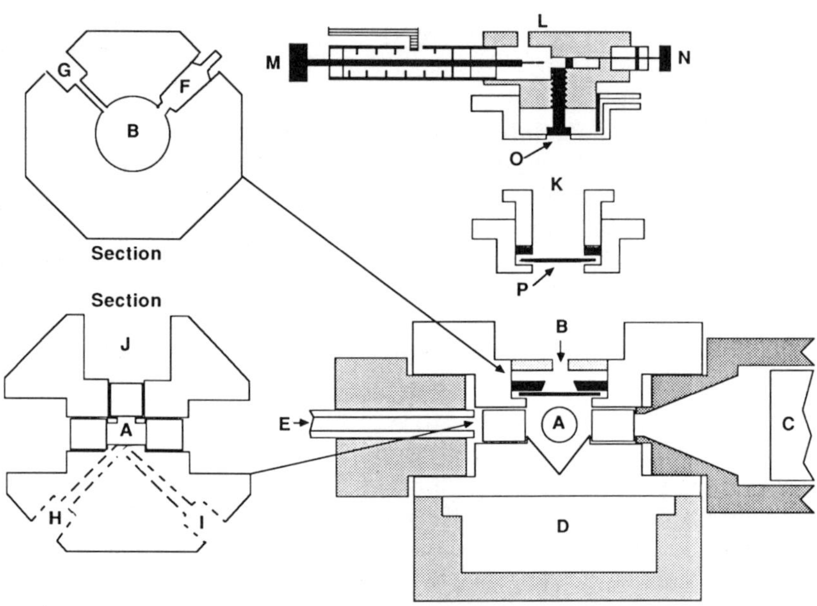

Figure 4.8. Schematic representation of the pressure-jump apparatus of Davis and Gutfreund (1976). The instrument is composed of the following components: A, observation cell; B, hydraulic chamber; C, absorbancy photomultiplier; D, thermostatted base; E, quartz fiber optic from light source; F, quartz pressure transducer for the triggering of data collection; G, hydraulic pressure line; H and I, observation cell filling and emptying ports; J, fluorescence emission window; K, bursting disc pressure-release valve; L, mechanical pressure-release valve; M, trigger mechanism; N, reset mechanism; O, value seat; and P, phosphorbronze bursting disc. (Reprinted with permission of the publisher.)

optimized S-20 type photomultiplier is used for optical detection and a 12-bit A/D converter interfaces the pressure jump unit to a computer. Hi-Tech offers two software packages, one for the Apple II e/IIgs and the other for the Hewlett-Packard 310 series technical desktop computer.

Application of Pressure-Jump Relaxation Techniques to Soil Constituents

Soil chemists have not yet applied p-jump techniques to the study of fast catalytic and exchange reactions on soil constituents. Perhaps this is due to a lack of knowledge of p-jump relaxation and, until recently, to the unavailability of commercial p-jump spectrometers.

Chemists have applied p-jump techniques to study a wide array of liquid and solid reactions, which are discussed in detail in two excellent books entitled "Chemical and Biological Applications of Relaxation Spectrometry" (Wyn-Jones, 1974) and "Techniques and Applications of Fast Reactions in Solution" (Gettins and Wyn-Jones, 1979).

The application of p-jump relaxation spectrometry to soil constituents has been investigated in a number of papers by Japanese chemists (see, e.g., Yasunaga and Ikeda, 1986) as well as by others (Hayes and Leckie, 1986). Studies have included adsorption/desorption kinetics at the TiO_2–H_2O interface (Ashida *et al.*, 1978), kinetics of TiO_2/IO_3 interactions on γ-Al_2O_3 (Hachiya *et al.*, 1980), proton adsorption/desorption mechanisms on Fe oxides (Astumian *et al.*, 1981), kinetics of hydrolysis of zeolites (Ikeda *et al.*, 1981), adsorption/desorption dynamics of acetic acid on silica–alumina particles (Ikeda *et al.*, 1982b), ion exchange kinetics of metals (Ikeda *et al.*, 1982a), and NH_4^+ adsorption on zeolites (Ikeda *et al.*, 1984b). A number of other adsorption and/or desorption kinetic studies all on γ-Al_2O_3 have included interactions with Pb^{2+} (Hachiya *et al.*, 1979; Hayes and Leckie, 1986), phosphate (Mikami *et al.*, 1983a), chromate (Mikami *et al.*, 1983b), and other metals (Hachiya *et al.*, 1984a,b). Recently, the Japanese group has applied p-jump relaxation to the study of intercalation catalytic reactions between metals and clay minerals and TiS_2 (Ikeda and Yasunaga, 1984; Negishi *et al.*, 1984; Sasaki *et al.*, 1985).

The above studies have clearly shown that p-jump relaxation measures *chemical kinetics* and thus one derives the actual rate constants. The implications these types of measurements have for ascertaining mechanisms of ion exchange and of catalytic reactions on soil constituents is tremendous. The application of p-jump relaxation to studying ion exchange kinetics of NH_4^+ on zeolite (Ikeda *et al.*, 1984b) and the adsorption/desorption of Pb^{2+} on γ-Al_2O_3 (Hachiya *et al.*, 1979) is presented below,

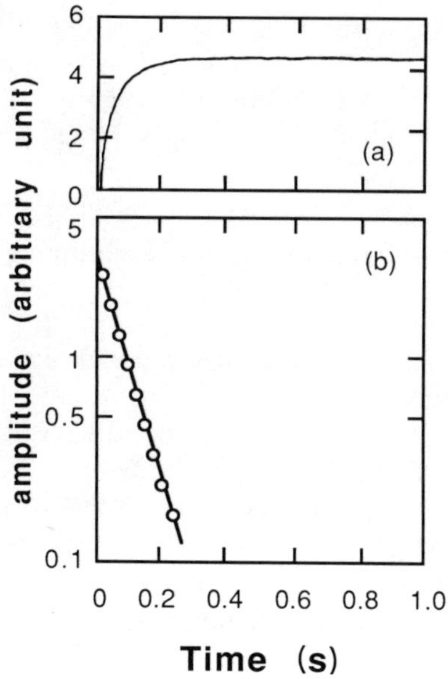

Figure 4.9. (a) Typical relaxation curve observed by using the pressure-jump relaxation method with electric conductivity detection in the H-ZSM-5-NH_4^+ system at the particle concentration of 20 g dm^{-3} at 298 K. (b) Semilogarithmic plot of the typical relaxation curve. [From Ikeda *et al.* (1984b), with permission.]

and shows how one can derive mechanistic data from p-jump studies.

Kinetics of Ion Exchange of NH_4^+ on Zeolite. Figure 4.9(a) shows a typical relaxation curve for aqueous suspensions of zeolite H-ZSM-5 and NH_4^+ using p-jump relaxation with conductivity detection, taken from the work of Ikeda *et al.*, (1984b). The direction of the relaxation signal increases during relaxation, and a semilogarithmic plot of the relaxation curve shows a single relaxation process [Fig. 4.9(b)]. The τ^{-1} values were determined at different concentrations of NH_4^+, and one sees that as $[NH_4^+]$ increases, τ^{-1} also increases (Fig. 4.10).

Ikeda *et al.* (1984b) proposed the following mechanism for ion exchange of NH_4^+ on the zeolite (S):

$$S(H) + NH_4^+ \underset{k_{-1}}{\overset{k_1}{\rightleftharpoons}} S(NH_4) + H^+ \tag{4.40}$$

The corresponding K_{eq} for the above reaction would be (Ikeda *et al.*, 1984b)

$$K_{eq} = \frac{[S(NH_4)][H^+]}{[S(H)][NH_4^+]} \tag{4.41}$$

where S(H) and $S(NH_4)$ denote the adsorbed sites of H^+ and NH_4^+. One

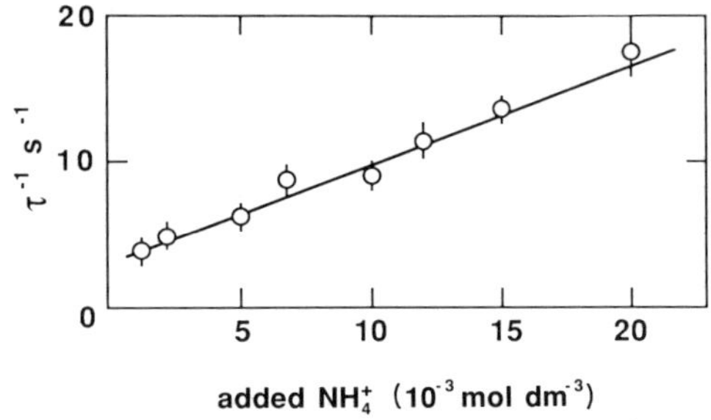

Figure 4.10. Dependence of the reciprocal relaxation time (τ^{-1}) on the concentration of added ammonium (NH_4^+) at the particle concentration of 20 g dm^{-3} at 298 K. [From Ideda *et al.*, (1984b), with permission.]

could then express τ^{-1} for Eq. (4.40) as

$$\tau^{-1} = k_1\{[S(H)] + [NH_4^+]\} + k_{-1}\{[S(NH_4)] + [H^+]\} \tag{4.42}$$

Ikeda *et al.* (1984b) plotted Eq. (4.42) by determining the equilibrium concentrations from adsorption isotherms for S(H), S(NH_4^+), and NH_4^+, and using the pH value to determine [H^+]. This plot shows good linearity (Fig. 4.11), which confirms that the mechanism hypothesized in Eq. (4.40) is operational. The k_1 and k_{-1} values for Eq. (4.42) can then be calculated from the slope and intercept of Fig. 4.11, and the kinetic K_{eq} can be determined from the ratio k_1/k_{-1} (Table 4.2). It is important to notice that the values calculated kinetically and statically (equilibrium method) are similar, which indicates that the rate constants one calculates from p-jump experiments are *chemical kinetics* rate constants. These data also verify

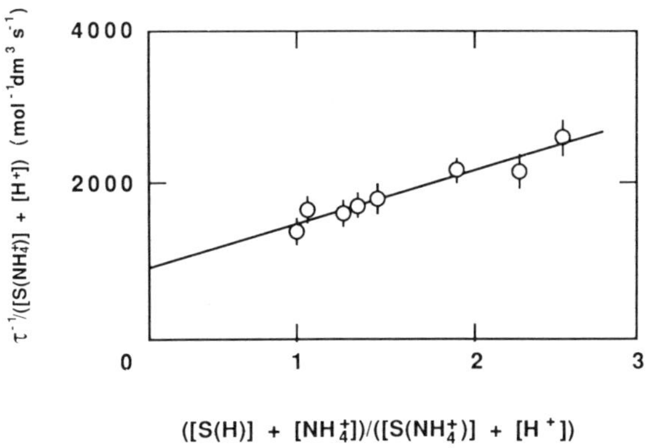

Figure 4.11. Plot of $\tau^{-1}/([S(NH_4)] + [H^+])$ versus $([S(H)] + [NH_4^+])/([S(NH_4)] + [H^+])$ in Eq. (4.42). [From Ikeda *et al.* (1984b), with permission.]

TABLE 4.2 Rate Constants of Ion Exchange of NH_4^+ for H^+ in Aqueous Suspensions of H-ZSM-5 at 298 K

		K_{eq}	
k_1 (mol^{-1} dm^3 s^{-1})	k_{-1} (mol^{-1} dm^3 s^{-1})	Kinetic[a]	Static
650	800	0.81	0.94

[a]This value was calculated from the ratio of the obtained rate constants k_1 and k_{-1}. From Ikeda *et al.* (1984b), with permission.

that the single relaxation that was observed in Fig. 4.9(b) by Ikeda *et al.* (1984b) could be ascribed to the ion exchange mechanism of NH_4^+ for H^+ in H-ZSM-5 zeolite given in Eq. (4.40).

Other evidence that would strongly suggest that the rate constants measured by p-jump relaxation are indeed chemical kinetics rate constants was given in the work of Ikeda *et al.* (1981). In this study, the kinetics of hydrolysis of zeolite 4A surface using p-jump relaxation and conductivity detection was determined. The τ^{-1} could be expressed as

$$\tau^{-1} = k_1([\text{SOH}] + [\text{OH}^-]) + k'_{-1} \tag{4.43}$$

with

$$k'_{-1} = k_{-1}[\text{H}_2\text{O}] \tag{4.44}$$

When τ^{-1} versus ([SOH] + [OH^-]) was plotted, a linear relationship resulted and k_1 and k'_{-1} were calculated as 1.6×10^2 mol^{-1} dm^3 s^{-1} and 8.7×10^{-2} s^{-1}, respectively. The average pK value calculated kinetically [p$K = -\log(k_1/k'_{-1})$] was -3.26 and was in excellent agreement with the average pK' value of -3.25 calculated from the equilibrium data.

Kinetics and Mechanisms of Adsorption-Desorption of Pb^{2+} *on* $\gamma-Al_2O_3$

Theoretical Considerations. Hachiya *et al.* (1979), who studied adorption–desorption kinetics of Pb^{2+} on a $\gamma-Al_2O_3$, proposed the following adsorption–desorption reaction scheme:

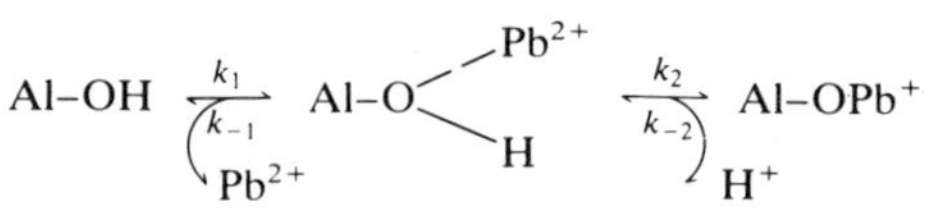

(4.45)

Step 1 τ_1 (fast process) Step 2 τ_2 (slow process) where step 1 is the adsorption–desorption of Pb^{2+} on the hydrous oxide surface group Al–OH, step 2 is the deprotonation–protonation induced by adsorbed Pb^{2+}, Al–O(H)—Pb^{2+} is the encounter complex formed by the adsorption of Pb^{2+}, Al–OPb^+ is the surface complex, τ_1 and τ_2 are the relaxation times of steps 1 and 2, respectively, and the k_i are the rate constants.

At constant ionic strength, Hachiya *et al.* (1979) gave the following rate equations for the two steps:

$$-d[Pb^{2+}]/dt = k_1 C_p(\Gamma_t^\infty - \Gamma_t)[Pb^{2+}] - k_{-1} C_p \Gamma_1 \qquad (4.45a)$$

$$-d[H^+]/dt = k_{-2} C_p \Gamma_2 [H^+] - k_2 C_p \Gamma_1 \qquad (4.45b)$$

with

$$K_1 = k_1/k_{-1} = \frac{\Gamma_1}{(\Gamma_t^\infty - \Gamma_t)[Pb^{2+}]} \qquad (4.46a)$$

$$K_2 = k_2/k_{-2} = \Gamma_2[H^+]/\Gamma_1 \qquad (4.46b)$$

and

$$\Gamma_t = \Gamma_1 + \Gamma_2 \qquad (4.46c)$$

where the square brackets indicate bulk concentrations; C_p the concentration of the particle; K the equilibrium constant; t, time; Γ, Γ^∞, and Γ_t the equilibrium, saturated, and total amount of adsorbed Pb^{2+}, respectively; and subscripts 1 and 2 refer to states Al–(OH)—Pb^{2+} and Al–OPb^+, respectively.

Perturbation of the equilibrium system results in alterations of bulk and surface concentrations of each species (Hachiya *et al.*, 1979):

$$[Pb^{2+}] = [\overline{Pb}^{2+}] + \Delta[Pb^{2+}] \qquad (4.47a)$$

$$[H^+] = [\bar{H}^+] + \Delta[H^+] \qquad (4.47b)$$

$$\Gamma_i = \bar{\Gamma}_i + \Delta\Gamma_i \qquad (i = 1, 2) \qquad (4.47c)$$

where the bar over the concentration terms indicates equilibrium values given by the law of mass conservation. Then

$$\Delta[Pb^{2+}] + C_p(\Delta\Gamma_1 + \Delta\Gamma_2) = 0 \qquad (4.48a)$$

$$\Delta[H^+] = C_p \, \Delta\Gamma_2 \qquad (4.48b)$$

With Eqs. (4.46)–(4.48), Eqs. (4.45) can be rewritten as

$$-d\,\Delta[Pb^{2+}]/dt = a_{11}\,\Delta[Pb^{2+}] + a_{12}\,\Delta[H^+] \qquad (4.49a)$$

$$-d\,\Delta[H^+]/dt = a_{21}\,\Delta[Pb^{2+}] + a_{22}\,\Delta[H^+] \qquad (4.49b)$$

with

$$a_{11} = k_1\{C_p(\Gamma_t^\infty - \bar{\Gamma}_t) + [\overline{Pb^{2+}}]\} + k_{-1}$$
$$a_{12} = k_{-1}$$
$$a_{21} = k_2$$
$$a_{22} = k_{-2}(C_p\bar{\Gamma}_2 + [\bar{H}^+]) + k_2$$

Accordingly, the values for τ^{-1} can be written as

$$\tau_{1,2}^{-1} = \tfrac{1}{2}\{(a_{11} + a_{22}) \pm [(a_{11} + a_{22})^2 - 4(a_{11}a_{22} - a_{12}a_{21})]^{1/2}\} \quad (4.50)$$

Equation (4.50) can be simplified under special conditions; for (*i*) $\tau_1^{-1} \ll \tau_2^{-1}$ ($a_{11} \ll a_{22}$),

$$\tau^{-1} = a_{11} - \frac{a_{12}a_{21}}{a_{22}}$$
$$= k_1\{C_p(\Gamma_t^\infty - \bar{\Gamma}_t) + [Pb^{2+}]\} + k_{-1} - \frac{k_{-1}k_2}{k_{-2}(C_p\bar{\Gamma}_2 + [\bar{H}^+] + k_2)} \quad (4.51a)$$

$$\tau_2^{-1} = a_{22} = k_{-2}(C_p\bar{\Gamma}_2 + [\bar{H}^+]) + k_2 \quad (4.51b)$$

For (*ii*) $\tau_1^{-1} \gg \tau_2^{-1}$ ($a_{11} \gg a_{22}$),

$$\tau_1^{-1} = a_{11} = k_1\{C_p(\Gamma_t^\infty - \bar{\Gamma}_t) + [\overline{Pb^{2+}}]\} + k_{-1} \quad (4.52a)$$

$$\tau_2^{-1} = a_{22} - \frac{a_{12}a_{21}}{a_{11}}$$
$$= k_{-2}(C_p\bar{\Gamma}_2 + [H^+]) + k_2 - \frac{k_{-1}k_2}{k_1\{C_p(\Gamma_t^\infty - \bar{\Gamma}_t) + [\overline{Pb^{2+}}]\} + k_{-1}} \quad (4.52b)$$

At equilibrium, the equation for the quantity of adsorbed Pb^{2+} is obtained from Eqs. (4.46a)–(4.46c):

$$\bar{\Gamma}_t^{-1} = (\Gamma_t^\infty)^{-1} + (\Gamma_t^\infty K_L)^{-1}[\overline{Pb^{2+}}]^{-1} \quad (4.53)$$

with $K_L = K_1(1 + K_2/[\bar{H}^+])$, where K_L is the pseudo-Langmuir constant. This equation is the same as the Langmuir isotherm equation if the value of $[H^+]$ is constant. The apparent and true overall equilibrium constant, $K_{o,ap}$ and K_{eq}, respectively, are related by the following equation:

$$K_{o,ap} \equiv \frac{\bar{\Gamma}_t[H^+]}{(\Gamma_t^\infty - \bar{\Gamma}_t)[\overline{Pb^{2+}}]} = K_o + K_1[\bar{H}^+] \quad (4.54)$$

with $K_{eq} = K_1K_2$.

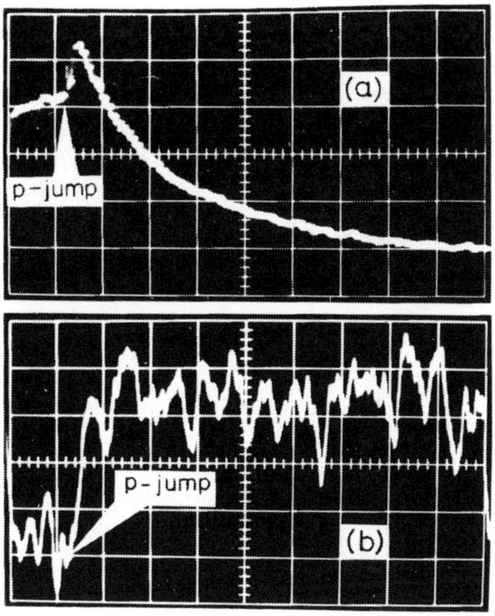

Figure 4.12. Typical relaxation curves in aqueous γ-Al_2O_3–$Pb(NO_3)_2$ suspension observed by the pressure-jump method with (a) electric conductivity and (b) turbidity detection. Concentration of Al_2O_3, C_p, is 15 g dm^{-3} at 293 K; sweep, 2 ms/division; wavelength in (b), 525 nm. [From Hachiya *et al.*, 1979), with permission.]

Mechanisms of Pb^{2+} Adsorption-Desorption. Hachiya *et al.* (1979) used the previous theoretical development considerations to test a number of mechanisms for Pb^{2+} reactions on a γ-Al_2O_3. They made measurements on an aqueous suspension of γ-Al_2O_3 containing $Pb(NO_3)_2$ at 293 K. In Fig. 4.12a one sees a typical p-jump relaxation curve using conductivity detection. The curve increases with pressure, and initially a rapid conductivity change is observed. Hachiya *et al.* (1979) also carried out other experiments using p-jump and the electric field pulse technique (not shown) in a γ-Al_2O_3 suspension, an aqueous solution of $Pb(NO_3)_2$, and a γ-Al_2O_3–$NaNO_3$ suspension. However, no relaxation effects were observed. Since the Al_2O_3-$Pb(NO_3)_2$ suspension was slightly acidic, the authors conducted an experiment in a γ-Al_2O_3 suspension to ascertain whether protons contributed to relaxation, but no relaxation was found. A relaxation similar to that in the γ-Al_2O_3-$Pb(NO_3)_2$ suspension was observed in a γ-Al_2O_3-$PbCl_2$ system. From semi-log plots of the relaxation curves in Fig. (4.12), it was determined that two kinds of relaxations occurred in the γ-Al_2O_3 suspension with Pb^{2+}. The authors suggested that the observed relaxations (after experimentally eliminating the possibility of aggregation–dispersion of γ-Al_2O_3 particles) were caused by Pb^{2+} adsorption–desorption phenomena on the γ-Al_2O_3 surface.

Hachiya *et al.* (1979) determined the relationship between slow and fast reciprocal relaxation times (τ_s^{-1} and τ_f^{-1}, respectively) at various ionic strength and pH values and found that the value of τ_s^{-1} depended on the concentration of Pb^{2+} and pH but was independent of ionic strength, whereas τ_f^{-1} decreased with increasing Pb^{2+} concentration.

To explain why τ_s^{-1} depended on Pb^{2+} concentration, Hachiya *et al.* (1979) examined five possible mechanisms that have been suggested in the literature and which are elaborated on below.

Mechanism I: Langmuir Type Adsorption–Desorption Reaction. Consider the reaction

$$\text{Al–OH} + \text{Pb}^{2+} \underset{k_{-1}}{\overset{k_1}{\rightleftharpoons}} \text{Al–O}\genfrac{}{}{0pt}{}{\diagup \text{Pb}^{2+}}{\diagdown \text{H}} \tag{4.55}$$

The authors simplified the rate equation for such a reaction so that the zeta (ζ) potential on the surface of γ-Al_2O_3 particles was nearly constant. Thus,

$$-d[\text{Pb}^{2+}]/dt = k_1\{C_P(\Gamma^\infty - \Gamma) + [\text{Pb}^{2+}]\} + k_{-1}C_p\Gamma \tag{4.56}$$

with $K_{eq} = k_1/k_{-1}$ where Γ is the quantity of Al–O(H)-–Pb^{2+} and $(\Gamma^\infty - \Gamma)$ is the effective concentration of Al–OH on the surface. If Eq. (4.56) is solved, an expression for τ^{-1} can be given as

$$\tau^{-1} = k_1\{C_p(\Gamma^\infty - \bar{\Gamma}) + [\overline{\text{Pb}}^{2+}] + K_{eq}^{-1}\} \tag{4.57}$$

Hachiya *et al.* (1979) investigated the relationship between τ_s^{-1} and the expression in braces in Eq. (4.57), using experimental data but a nonlinear relationship was observed. Based on this nonlinearity, mechanism I was excluded.

Mechanism II: Adsorption–Desorption Reaction Involving Ion Exchange with Two Protons. This mechanism was investigated similarly to mechanism I but did not appear to be operational.

Mechanism III: Adsorption–Desorption Reaction Involving Ion Exchange with Proton. This reaction for the slow process can be given as (Hachiya *et al.*, 1979)

$$\text{Al–OH} + \text{Pb}^{2+} \rightleftharpoons \text{Al–OPb}^{+} + \text{H}^{+} \tag{4.58}$$

and the rate equation for mechanism III would be

$$-d[\text{Pb}^{2+}]/dt = k_1C_p(\Gamma^\infty - \Gamma)[\text{Pb}^{2+}] - k_{-1}C_p\Gamma[\text{H}^+] \tag{4.59}$$

where Γ denotes the amount of Al–OPb^+.

The equation for τ^{-1} would be

$$\tau^{-1} = k_{-1}[C_p\bar{\Gamma} + [\text{H}^+] + K_{o,ap}\{C_p(\Gamma^\infty - \bar{\Gamma}) + [\overline{\text{Pb}}^{2+}]\} \tag{4.60}$$

with $K_{o,ap} = k_1/k_{-1}$.

Experimental plots of τ_s^{-1} versus the terms between brackets in Eq. (4.60) did not result in linear relationships, so mechanism III was also ruled out by Hachiya *et al.* (1979).

Mechanism IV: Adsorption–Desorption Reaction in the Presence of a Proton Catalyst. The mechanism in which H^+ acts as a catalyst is

$$\text{Al–OH} + Pb^{2+} + H^+ \rightleftharpoons \text{Al–OPb}^+ + 2H^+ \tag{4.61}$$

This mechanism was examined by Hachiya *et al.* (1979) but did not appear to explain the dependence of τ_s^{-1} on concentration.

Mechanism V: Two-Step Adsorption–Desorption Reactions Involving Ion Exchange with a Proton. Hachiya *et al.* (1979) felt that the existence of two relaxations in Fig. 4.12 suggested that the Pb^{2+} adsorption–desorption reactions consisted of two processes, one fast and the other slow.

Mechanism III could thus be further divided into two elementary steps:

$$\text{(Va)} \quad \text{Al–OH} \xrightarrow[H^+]{} \text{Al–O}^- \xrightarrow[Pb^{2+}]{} \text{Al–OPb}^+ \tag{4.62}$$

Step 1 Step 2

In the case (Vb), the mechanism given by Eq. (4.45), step 1 is very fast compared to step 2 in mechanism Va. Or, in the converse case, the value of τ_s^{-1} calculated from the rate equations for the two steps is not proportional to the amount of adsorbed Pb^{2+}.

Hachiya *et al.* (1979) then examined the experimental results for τ_s^{-1} using Eqs. (4.51a) and (4.52b) for mechanism Vb. With Eq. (4.51a) the value of τ_s^{-1} should not be proportional to the amount of adsorbed Pb^{2+}, but should be according to Eq. (4.52b). As a result, for the slow relaxation, Eq. (4.52b) can be deduced as (Hachiya *et al.*, 1979)

$$\tau_s^{-1} \simeq k_{-2} \frac{C_p\bar{\Gamma}t}{1 + K_2^{-1}[\bar{H}^+]} \tag{4.63}$$

The value of K_2 can be ascertained from the slope and intercept of the straight line shown in Fig. 4.13 where the dependence of $K_{o,ap}$ on $[H^+]$ is given by Eq. (4.54). The value for Γ_t^∞ in Eq. (4.54) can be found from Eq. (4.53). The experimental plots of τ_s^{-1} versus $C_p\bar{\Gamma}_t(1 + K_2^{-1}[\bar{H}^+])^{-1}$ show a linear relationship through the origin (Fig. 4.14). Hachiya *et al.* (1979) concluded that this possibly verifies that the slow reaction for Pb^{2+} adsorption/desorption on γ–Al_2O_3 can be ascribed to step 2 in mechanism Vb, given by Eq. (4.45). The value of k_{-2} was obtained from the slope of the straight line in Fig. 4.14, and the value of k_2 was calculated using K_2 and k_2 values. These results are given in Table 4.3.

The mechanism for the fast process in Eq. (4.45) was also inferred by Hachiya *et al.* (1979). The Pb^{2+} concentration dependence of τ_f^{-1} versus the expression between braces in Eq. (4.52a) is shown in Fig. (4.15). A

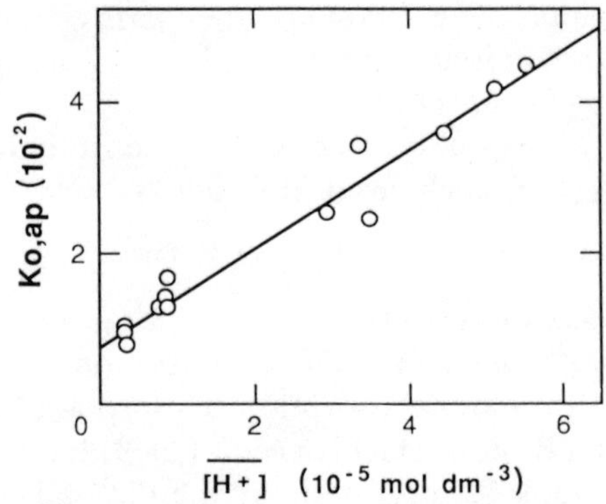

Figure 4.13. Plot of $K_{o'ap}$ versus $[\overline{H^+}]$. [From Hachiya *et al.* (1979), with permission.]

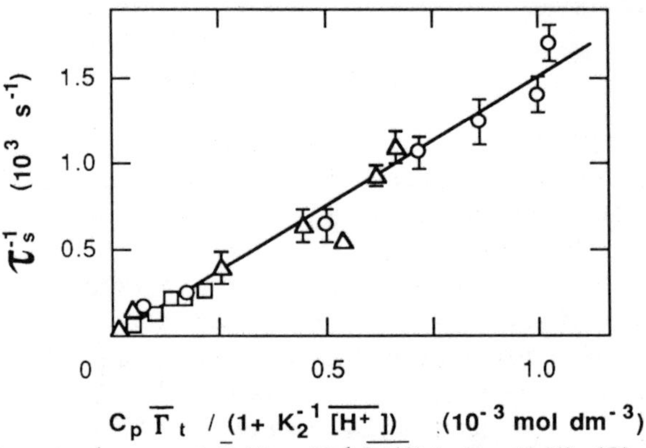

Figure 4.14 Plot of τ_s^{-1} versus $Cp\overline{\Gamma}_t/(1 + K_2^{-1}\ \overline{[H^+]})$ in Eq. (4.63); (○) pH 5.4, (N) pH 5.0, (□) pH 4.5. [From Hachiya *et al.* (1979), with permission.]

TABLE 4.3 Kinetic Parameters of Adsorption–Desorption of Pb^{2+} on γ-Al_2O_3 Particles at 293 K[a]

Step 1: k_1, 10^8 mol^{-1} dm^3 s^{-1}	k_{-1}, 10^4 s^{-1}	K_1, 10^2 mol^{-1} dm^3
1.4 ± 0.4	1.0 ± 0.4	14 ± 10 6.7 ± 0.3[b]
Step 2: k_2, 10s^{-1}	**k_{-2}, 10^6 mol^{-1} dm^3 s^{-1}**	**K_2, 10^{-6} mol dm^{-3}**
1.3 ± 0.6	1.5 ± 0.1	9 ± 4

[a] From Hachiya *et al.* (1979). with permission.

[b] This value was obtained from the slope of the straight line in Fig. 4.13.

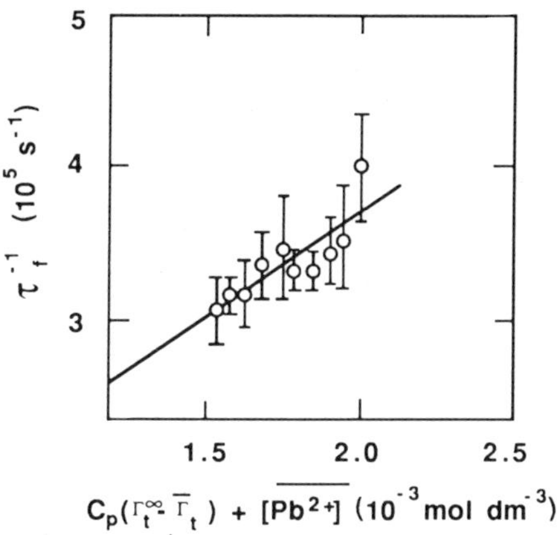

Figure 4.15. Plot of τ_f^{-1} versus the expression between braces in Eq. (4.52a). [From Hachiya *et al.* (1979), with permission.]

linear relationship results, and k_1 and k_{-1} values were calculated from the slope and intercept, respectively, of Fig. 4.15 and are given in Table 4.3. The value of K_1 calculated kinetically $(14 \pm 10) \times 10^2$ mol^{-1} dm^3 (Table 4.3) was in agreement with the K_1 value using Eq. (4.54). Thus, Hachiya *et al.* (1979) attributed the fast relaxation to step 1 in mechanism (Vb) given by Eq. (4.45).

STOPPED-FLOW TECHNIQUES

Introduction

Flow methods have also been widely used to measure rapid reaction rates in solution. They are forms of mix and shake methods that initiate reactions and that are concerned with fast mixing of the reactant and subsequently with rapid measurements of solution concentration changes (Robinson, 1986).

A rapid solution reaction begun by mixing two reagents can be observed two ways. The mixed reagents can be observed downstream from the mixing chamber, and, if flow continues the reaction mixture composition does not change with time. This method can be referred to as continuous flow. It requires a large quantity of reagents, but rapid detection methods are not necessary to follow the reaction. The second method is to study a stationary portion of the reactant mixture with a detection system that is fast enough to follow the reaction as it proceeds in the mixture. This method is known as stopped flow (Robinson, 1986).

The stopped-flow method is more often used than any other technique for observing fast reactions with half-lives of a few milliseconds. Another attribute of this method is that small amounts of reactants are used. One must realize, however, that flow techniques are relaxation procedures that involve concentration jumps after mixing. Thus, the mixing or perturbation time determines the fastest possible rate that can be measured. Stopped-flow methods have been widely used to study organic and inorganic chemical reactions and to elucidate enzymatic processes in biochemistry (Robinson, 1975; 1986). The application of stopped-flow methods to study reactions on soil constituents is very limited to date (Ikeda *et al.*, 1984a).

Stopped-Flow Instrumentation and Design

For those primarily interested in the design and construction of stopped-flow instruments, the reader should consult contributions by a number of authors (Caldin, 1964; Gibson, 1969; Kustin, 1969; Chance, 1974). Commercial stopped-flow instruments are readily available and are manufactured in the United States, England, France, and Japan. Two of the current commercial units are discussed below. The major advance in stopped-flow instruments since 1970 has been in adapting stopped-flow units to a variety of detection systems, which has increased the kinds of reactions that can be studied. There is also stopped-flow instrumentation that enables reaction rates to be studied as a function of pressure.

Hi-Tech Scientific Limited (Salisbury, England) recently introduced a stopped-flow (SF-51) instrument with conductivity detection that uses a five-mixer aging block that gives preparative quench aging times in the range of 1.0 ms to $\geq$10 s (Fig. 4.16). Preparative quench and stopped-flow experiments can be performed under total thermostatted, anaerobic, and chemically inert conditions. The entire stopped-flow package consists of the sample handling unit, a spectrophotometer, and a data processor based on the Apple IIe.

Bio-Logic Instrument and Laboratories (Meylan, France) manufactures an SFM-3 stopped-flow instrument (Fig. 4.17) that consists of three independent drive syringes driven by stepping motors, two mixers and a delay line, three observation windows, replaceable cuvettes, no stop-syringes, and efficient temperature regulation. At maximum flow rate, the minimum dead times range from 1.0 to 4.9 ms for fluorescence detection and 1.3 ms for transmittance. Currently, the Bio-Logic MOS-1000 optical system employs fluorescence or absorbance detection, which is not suitable

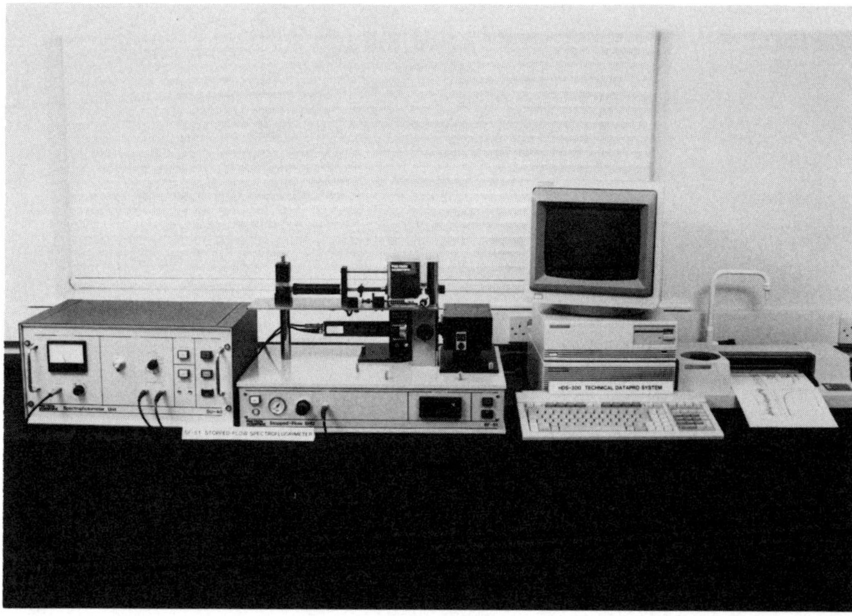

Figure 4.16. Stopped-flow apparatus (SF-51) from Hi-Tech Scientific Limited with datapro system. (Used with permission of the manufacturer.)

for soil constituent kinetic studies, but a conductivity detection is being introduced soon. The SFM-3 module is controlled by the user from the keyboard of a PC XT or compatible microcomputer. One screen allows the user to enter the drive and aging sequences, to perform manual or automatic movement of the syringes, and to save or recall the drive sequences. It also displays the volumes of the solutions contained in the three syringes. The same microcomputer may also be used as a data acquisition system with the Bio-Kine software package. This package is based on an IBM-PC XT/AT or other compatible computer and provides all software and hardware necessary for the acquisition and treatment of rapid kinetic data. Data treatment includes visualization of recorded files, background subtraction, multiexponential analysis, mathematical operations between files, and hard copy or plot.

Application of Stopped-Flow Techniques to Soil Constituent Reactions

At this writing, the only application of stopped-flow methodology to soil constituent dynamics has involved studies on the cation exchange kinetics of Li^+, K^+, Rb^+, and Cs^+ for Na^+ on a zeolite (Ikeda *et al.*, 1984a).

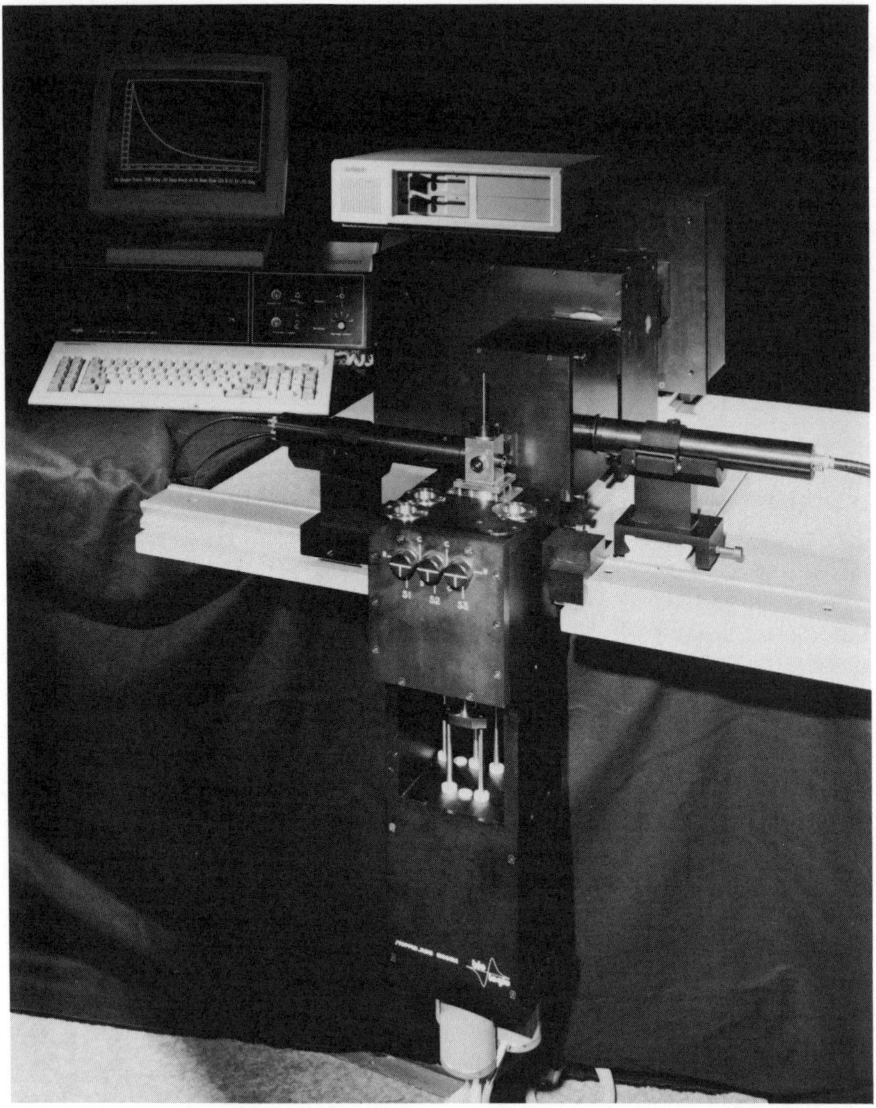

Figure 4.17. Stopped flow apparatus (SFM-3) of Bio-Logic Instruments and Laboratories. (Used with permission of the manufacturer.)

In the latter study, an aqueous solution of the alkali metal ion in excess was mixed with an aqueous suspension of zeolite particles ($<1\ \mu m$). The reaction of the mixed suspension was observed using conductivity detection with a dead time of 15 ms. Ikeda *et al.* (1984a) observed a conductivity

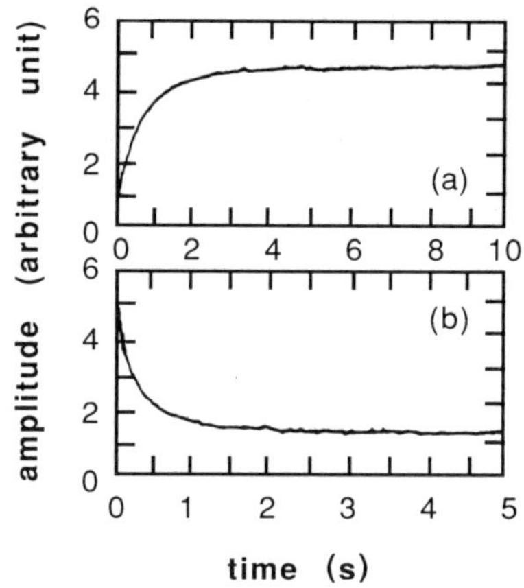

Figure 4.18. Typical reaction curves observed by using the stopped-flow method with electrical conductivity detection at Cp = 0.7 g dm^{-3} and 298 K: (a) Li^+, (b) K^+, Rb^+, and Cs^+. [From Ikeda *et al.* (1984a), with permission.]

increase with time for the Li^+–Na^+ system and a conductivity decrease with time for the K^+–, Rb^+–, and Cs^+ –Na^+ systems (Fig. 4.18). Based on kinetic and equilibrium studies, the authors attributed the former observation to the release of Na^+ induced by the Li^+ adsorption and the latter to the adsorption of K^+, Rb^+, or Cs^+ on the site in the cage of zeolite 4A.

ELECTRIC FIELD METHODS

Introduction

Another chemical relaxation method that can be used to determine the kinetics of fast reactions on soil constituents is the electric field pulse technique. This technique was developed by Hachiya *et al.* (1980) to study the kinetics of IO_3^- adsorption and desorption on TiO_2 and by Sasaki *et al.* (1983) to investigate ion-pair formation on the surface of α-FeOOH. Excellent review articles on electric field methods are found in DeMaeyer (1969), Hemmes (1979), and Eyring and Hemmes (1986).

One can easily produce strong electric fields ($30\ kV/cm < E < 100\ kV/cm$) between metal electrodes by spacing them about 1 cm apart in a liquid. Such a pulsed field can result in reaction perturbation of an equilibrium in the liquid. Following the perturbation, a variety of methods can be utilized to determine the reaction kinetics (Eyring and Hemmes, 1986).

Application of the Electric Field Pulse Technique

Hachiya *et al.* (1980) and Sasaki *et al.* (1983) appear to be the only researchers who have applied electric field pulse techniques to the study of kinetics of soil constituent reactions. In the latter study, ion-pair formation

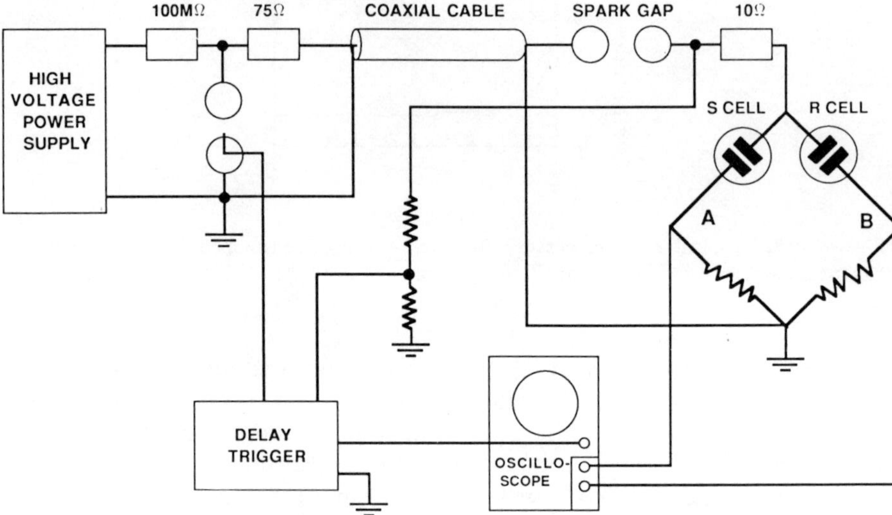

Figure 4.19. The block diagram of an electric field pulse apparatus. [From Ilgenfritz (1966), with permission.]

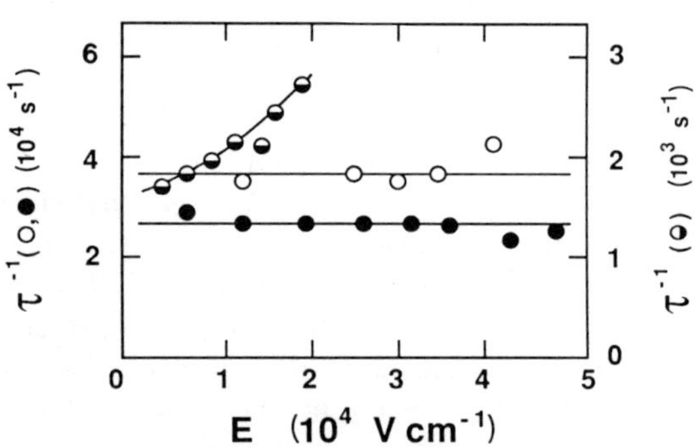

Figure 4.20. Electric field intensity dependence of the reciprocal fast and slow relaxation times in an acidic ∝-FeOOH suspension at 298 K. The reciprocal fast relaxation time in ∝-FeOOH-$HClO_4$ system (◒): Cp = 29.1 g dm^{-3}, [$HClO_4$] = 1.0×10^{-4} *M*, and salt-free. The reciprocal slow relaxation times in ∝-FeOOH–$HClO_4$ (○) and ∝-FeOOH–HCl (●) systems: Cp = 20 g dm^{-3}, [acid] = 1.75×10^{-3} *M*, and [salt] = 1.25×10^{-4} mol dm^{-3}. [From Sasaki *et al.* (1983), with permission.]

kinetics on α-FeOOH were studied using conductivity detection. A step-function electric field (rise time <0.1 μs) of 41 kV cm^{-1} was applied using an experimental setup originally constructed by Ilgenfritz (1966) and shown in Fig. (4.19). Two relaxations were observed over the pH range of 3.2–5.5 in acidic suspensions of α-FeOOH containing HCl and $HClO_4$. The rapid conductivity decreasing relaxation was observed only in the range of $[A^-] \lesssim 2 \times 10^4$ mol dm^{-3}, where $[A^-]$ represents the anion concentration, and is ascribed to a rapid polarization of the electrical double layer, followed by induced adsorption of ions. The τ^{-1} values were proportional to electric field strength (Fig. 4.20). The slower relaxation was observed as an increase in the conductivity of the electric field and was independent of the applied electric field intensity. This relaxation was ascribed to the dissociation field effect and was related to the association–dissociation reaction of counterions with the protonated surface hydroxyl groups (Sasaki *et al.*, 1983).

SUPPLEMENTARY READING

Bernasconi, C. F. (1976). "Relaxation Kinetics." Academic Press, New York.

Bernasconi, C. F., ed. (1986). "Investigations of Rates and Mechanisms of Reactions," 4th ed. Wiley, New York.

Chance, B. (1974). Rapid flow methods. *In* "Investigation of Rates and Mechanisms of Reactions" (G. G. Hammes, ed.), 3rd ed., pp. 187–210. Wiley, New York.

Davis, J. S., and Gutfreund, H. (1976). The scope of moderate pressure changes for kinetic and equilibrium studies of biochemical systems. *FEBS Lett.* **72,** 199–207.

Eigen, M., and DeMaeyer, L. (1963). Relaxation methods. *Tech. Org. Chem.* **8** (2), 895–1054.

Eyring, E. M., and Hammes, P. (1986). Electric field methods. *In* "Investigations of Rates and Mechanisms of Reactions" (C. F. Bernasconi, ed.), 4th ed., pp. 219–246. Wiley, New York.

Gettins, W. J., and Wyn-Jones, E., eds. (1979). "Techniques and Applications of Fast Reactions in Solution." Reidel Publ., Dordrecht, The Netherlands.

Gruenewald, B., and Knoche, W. (1979). Recent developments and applications of pressure jump methods. *In* "Techniques and Applications of Fast Reactions in Solutions" (W. J. Gettins and E. Wyn-Jones, eds.), pp. 87–94. Reidel Publ., Dordrecht, The Netherlands.

Hachiya, K., Ashida, M., Sasaki, M., Kan, H., Inoue, T., and Yasunaga, T. (1979). Study of the kinetics of adsorption-desorption of Pb^{2+} on a γ-Al_2O_3 surface by means of relaxation techniques. *J. Phys. Chem.* **83,** 1866–1871.

Hayes, K. F., and Leckie, J. O. (1986). Mechanism of lead ion adsorption at the goethite-water interface. *ACS Symp. Ser.* **323,** 114–141.

Hemmes, P. (1979). Electric field methods. *In* "Techniques and Applications of Fast Reactions in Solutions" (W. J. Gettins and E. Wyn-Jones, eds.), Vol. 50, pp. 95–101. Reidel Publ., Dordrecht, The Netherlands.

Ikeda, T., and Yasunaga, T. (1984). Kinetic studies of ion exchange of NH_4^+ for H^+ in zeolite H-ZSM-5 by the chemical relaxation method. *J. Colloid Interface Sci.* **99,** 183–186.

Ikeda, T., Sasaki, M., Astumian, R. D., and Yasunaga, T. (1981). Kinetics of the hydrolysis of zeolite 4A surface by the pressure-jump relaxation method. *Bull. Chem. Soc. Jpn.* **54,** 1885–1886.

Ikeda, T., Nakahara, J., Sasaki, M., and Yasunaga, T. (1984). Kinetic behavior of alkali metal ion on zeolite 4A surface using the stopped-flow method. *J. Colloid Interface Sci.* **97,** 278–283.

Ikeda, T., Sasaki, M., and Yasunaga, T. (1984). Kinetic studies of ion exchange of NH_4^+ in zeolite H-ZSM-5 by the chemical relaxation method. *J. Colloid Interface Sci.* **98,** 192–195.

Knoche, W. (1974). Pressure-jump methods. *In* "Investigations of Rates and Mechanisms of Reactions" (G. G. Hammes, ed.), 3rd ed., pp. 187–210. Wiley, New York.

Mikami, N., Sasaki, M., Hachiya, K., Astumian, R. D., Ikeda, T., and Yasunaga, T. (1983). Kinetics of the adsorption-desorption of PO_4 on the Al_2O_3 surface using the pressure-jump technique. *J. Phys. Chem.* **87,** 1454–1458.

Robinson, B. H. (1975). The stopped-flow and temperature-jump techniques-principles and recent advances. *In* "Chemical and Biological Applications of Relaxation Spectrometry" (E. Wyn-Jones, ed.), pp. 41–48. Reidel Publ., Dordrecht, The Netherlands.

Robinson, B. H. (1986). Rapid flow methods. *In* "Investigations of Rates and Mechanisms of Reactions" (C. F. Bernasconi, ed.), 4th ed., pp. 9–26. Wiley, New York.

Takahashi, M. T., and Alberty, R. A. (1969). The pressure-jump method. *In* "Methods in Enzymology" (K. Kustin, ed.), Vol. 16, pp. 31–55. Academic Press, New York.

Yasunaga, T., and Ikeda, T. (1986). Adsorption-desorption kinetics at the metal-oxide-solution interface studied by relaxation methods. *ACS Symp. Ser.* **323,** 230–253.

5

Ion Exchange Kinetics on Soils and Soil Constituents

INTRODUCTION

Ion exchange on soils and soil constituents has been known since Aristotle's time. It was rediscovered as base exchange by soil scientists in the nineteenth century (Thompson, 1850; Way, 1850), and since that time a voluminous amount of research has appeared in the scientific literature. However, most of these studies, particularly in the soil and environmental sciences, have dealt with equilibrium aspects of ion exchange, and only recently have studies appeared on the kinetics and mechanisms of exchange phenomena.

As mentioned in Chapter 1, J. Thomas Way (1850) was perhaps the first person to study rates of exchange on soils. Little else appeared in the literature on ion exchange kinetics until the work of Gedroiz (1914) and Hissink (1924). Some of the first studies on ion exchange kinetics assumed a *chemical reaction* (CR) that could be explained using rate coefficients and kinetic orders corresponding to exchange stoichiometry as shown by (Helfferich, 1983)

$$d\langle\bar{C}_1\rangle/dt = kC_1^{\nu_1}\langle\bar{C}_2\rangle^{\nu_2} - (k/\alpha_{12})C_2^{\nu_2}\langle\bar{C}_1\rangle^{\nu_1} \tag{5.1}$$

where $\langle\bar{C}_1\rangle$ and $\langle\bar{C}_2\rangle$ refer to average concentrations of ions 1 and 2, respectively, on the exchanger phase; C_1 and C_2 are concentrations of ions 1 and 2; t is time; k is the reaction rate coefficient; ν_1 and ν_2 are the stoichiometric coefficients for ions 1 and 2, respectively; and α_{12} is the separation factor for ions 1 and 2.

However, researchers later found (Boyd *et al.*, 1947) that the rate of ion exchange increased with decreasing particle size of the exchanger. This showed that mass-transfer phenomena and *not* chemical reaction were rate-controlling.

The first detailed study on ion exchange rates, and particularly mechanisms, appeared in the very definitive and elegant studies of Boyd *et al.* (1947) with zeolites. Working in conjunction with the Manhattan Project, these researchers clearly showed that ion exchange is diffusion-controlled, and that the reaction rate is limited by mass-transfer phenomena that are either film (FD) or particle (PD) diffusion-controlled. Boyd *et al.* (1947) were also the first to derive rate laws for FD, PD, and CR. Additionally, they demonstrated that particle size had no effect on reaction control, that in FD the rate was inversely proportional to particle size, and that the PD rate was inversely proportional to the square of the particle size.

Following the landmark discoveries of Boyd *et al.* (1947), Helfferich and co-workers (Helfferich, 1956, 1962b, 1963, 1965, 1966; Helfferich and Plesset, 1958; Plesset *et al.*, 1958; Schlogle and Helfferich, 1957) made numerous contributions to our understanding of ion exchange kinetics. These workers as well as others (Glaski and Dranoff, 1963; Smith and Dranoff, 1964; Sharma *et al.*, 1970) showed that since ions carry charge, they are influenced by the electric field their diffusion causes. These discoveries were instrumental in proving that ion exchange diffusion obeys the Nernst–Planck equations better than Fick's laws for diffusion. Helfferich (1983), commenting on the historical aspects of ion exchange kinetics, notes that this was actually a rediscovery since the Nernst–Planck equations were first formulated in 1913 to study ion diffusion in glasses and their use had been widely promulgated by membrane scientists.

FICKIAN AND NERNST–PLANCK DIFFUSION EQUATIONS

If the co-ion concentration in the ion exchanger is quite low, then the concentrations and fluxes of the exchanging counterions A and B are tightly coupled with each other (Helfferich, 1966). The total counterion concentration must remain constant and be equal to the concentration of fixed ionic groups such that

$$|z_A|\bar{C}_A + |z_B|\bar{C}_B = \bar{C} \tag{5.2}$$

where z_A and z_B are the valences of counterions A and B, respectively, $\bar{C}_A$ and $\bar{C}_B$ are the concentrations of counterions A and B, respectively, on the exchanger, and $\bar{C}$ is the total counterion concentration on the exchanger.

The fluxes of A and B in opposite directions have to be equal in magnitude so that no net transfer of electric charge occurs, such that

$$z_A\bar{J}_A + z_B\bar{J}_B = 0 \tag{5.3}$$

where $\bar{J}_A$ and $\bar{J}_B$ are the fluxes of counterions A and B on the exchanger. Since this is rigid coupling, only one flux equation, for either A or B, can be used to calculate the kinetic behavior.

Earlier theories proposed (Boyd *et al.*, 1947; Kressman and Kitchener, 1949) that Fick's first law with a *constant* interdiffusion coefficient could be utilized as the flux equation:

$$\bar{J}_A = -\bar{D}\ \text{grad}\ \bar{C}_A \qquad (\bar{D} = \text{constant}) \tag{5.4}$$

where $\bar{D}$ is the diffusion or interdiffusion coefficient on the exchanger.

Equation (5.4) is correct if the counterions, A and B, have the same mobility, that is, $\bar{D}_A = \bar{D}_B = \bar{D}$. However, the Nernst–Planck model, which takes into account the effect of electrical potential gradients, challenges Eq. (5.4) if counterions of different mobilities are present.

According to the Nernst–Planck model, the faster counterion usually diffuses more rapidly. This resulting small net transfer of electric charge builds up an electrical-potential gradient that affects both counterion species, transferring both in the direction of diffusion of the slower counterion. This transference slowers the faster ion net flux and increases that of the slower ion. Thus, electroneutrality is restored. This automatic generation of the electric field, which superimposes electric transference on transport by diffusion, is the mechanism by which the ion exchange system conforms to restrictions given in Eqs. (5.2) and (5.3). To quantitatively treat such a system, Fick's law [Eq. (5.4)] is replaced by the

Nernst–Planck equations, which contain another term for electric transference:

$$\bar{J}_A = -\bar{D}_A \text{ grad } \bar{C}_A - \bar{D}_A\bar{C}_A(z_A F/RT) \text{ grad } \Psi \tag{5.5}$$

where Ψ is the electrical potential. Now, if the Nernst–Planck equations for A and B are combined with Eqs. (5.2) and (5.3) the following equation results:

$$\bar{J}_A = -\bar{D} \text{ grad } \bar{C}_A \tag{5.6}$$

where

$$\bar{D} = \bar{D}_A\bar{D}_B(z_A^2\bar{C}_A + z_B^2\bar{C}_B)/(z_A^2\bar{C}_A\bar{D}_A + z_B^2\bar{C}_B\bar{D}_B) \tag{5.7}$$

The $\bar{D}$ is not constant but depends on the relative concentrations of A and B. With vanishing concentration of either ion, $\bar{D}$ approaches the diffusion coefficient of that ion.

The Nernst–Planck equations are only limiting laws for ideal systems. If activity coefficient gradients are present, an additional term in Eq. (5.5) is created such that

$$\bar{D} = \frac{\bar{D}_A\bar{D}_B z^2\bar{C}_A + z_B^2\bar{C}_B + z_A\bar{C}_A z_B\bar{C}_B\,[d \ln K_a/d\,(z_A\bar{C}_A)]}{z_A^2\bar{C}_A\bar{D}_A + z_B^2\bar{C}_B\bar{D}_B} \tag{5.8}$$

where K_a is the corrected molar selectivity coefficient, which equals $\bar{y}_B^{z_A}/\bar{y}_A^{z_B}$, where $\bar{y}$ is the partial molar activity coefficient on the exchanger and can be found through independent equilibrium measurements (Helfferich, 1956).

Even though the Nernst–Planck equations work well in many instances and their theoretical basis is sound, Helfferich (1983) mentions several cases where they are not statisfactory. For example, they may not work well in situations where other processes besides mass transfer occur. This could occur if ion mobilities increase or decrease such that diffusion coefficients in the Nernst–Planck equations are not constant and thus particle size of the ion exchanger is affected. In zeolites, for example, which are rigid, the exchange of one counterion for another of different size affects the ease of motion and creates diffusion coefficient differences.

Even where ion exchange is not affected by the above factors, the Nernst–Planck equations are not very useful for diffusion phenomena in the film. After all, the Nernst film is somewhat enigmatic and there is a combination of diffusive and convective mass transfer that changes from the bulk solution to the particle surface. Nernst (1904) originally defined the outer limit of film only as the point where the concentration profile, if linearly extrapolated from the particle surface, reaches the concentration level of the bulk solution.

Another problem with the Nernst–Planck equations for studying ion

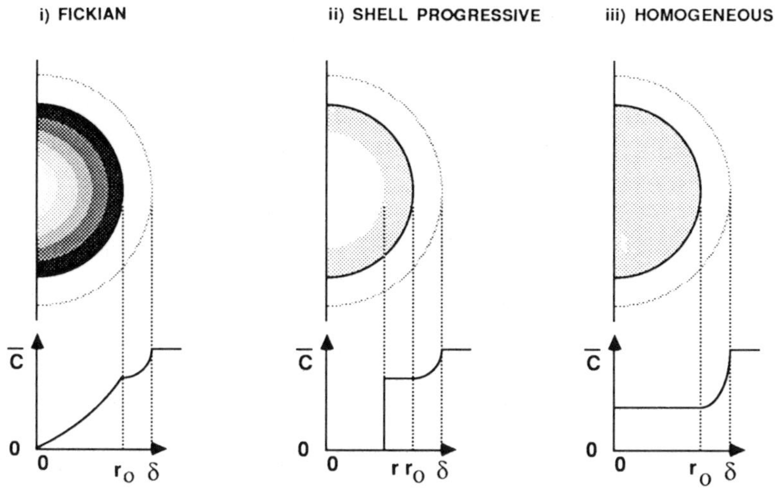

Figure 5.1. Diffusion mechanisms and corresponding concentration profiles, where $\bar{C}$ is the overall solid concentration, r_0 is the mean radius of the particle, r is the linear distance along the particle radius from the radius surface, and δ is the thickness of the stagnant fluid film. [From Liberti and Passino (1983), with permission.]

exchange kinetics was recently revealed in the work of Liberti and Passino (1985). By studying Cl/SO_4 exchange kinetics on anion exchange resins, it was found that ion selectivity can affect rate-limiting steps that may not be predicted by Nernst–Planck models. Depending on resin selectivity toward SO_4, homogeneous or shell-progressive diffusion may be taking place rather than Fickian diffusion (Fig. 5.1), which may not be described by earlier theories. Better theories are needed to look at the connection between selectivity and rate. Other techniques such as autoradiography, scanning electron microscopy, and X-ray microprobe analysis should be employed. With these, one could then look at the distribution of ions within particles, rather than concentration variations or model equations.

RATE-LIMITING STEPS

Concurrent Processes Involved in Ion Exchange

In an ion exchange kinetics study, any one or more of five steps can be rate-controlling. As an illustration of this, consider Na–K exchange on vermiculite (Sparks, 1986):

$$\text{Na} + \text{K-vermiculite} \rightleftharpoons \text{K} + \text{Na-vermiculite} \tag{5.9}$$

The co-ion in the above reactions is left out since usually it does not directly affect the rate of the reaction. However, it may affect the selectivity of the exchanger. Selectivity can affect rate-limiting steps in certain cases (Liberti and Passino, 1985). For the above reaction to take place, the following concurrent processes also take place: (1) diffusion of Na ions as Na and Cl through the solution film that surrounds the vermiculitic particles (FD); (2) diffusion of Na ions through a hydrated interlayer space of the vermiculite particle (PD); (3) chemical reaction exchange of Na ions for K ions from the particle surface (CR); (4) diffusion of displaced K ions through the hydrated interlayer space of the vermiculite particle (PD); and (5) diffusion of displaced K ions as K and Cl through the solution film away from the particle (FD).

Thus, ion exchange kinetics on heterogeneous surfaces such as vermiculite involve mass transfer (PD and FD) and CR processes (Fig. 5.2). For actual CR to occur, ions must be transported to the active fixed sites of the particles. The film of water adhering to and surrounding the particle and the hydrated interlayer spaces in the particle are both zones of low concentration. These zones are constantly being depleted by ion adsorption to

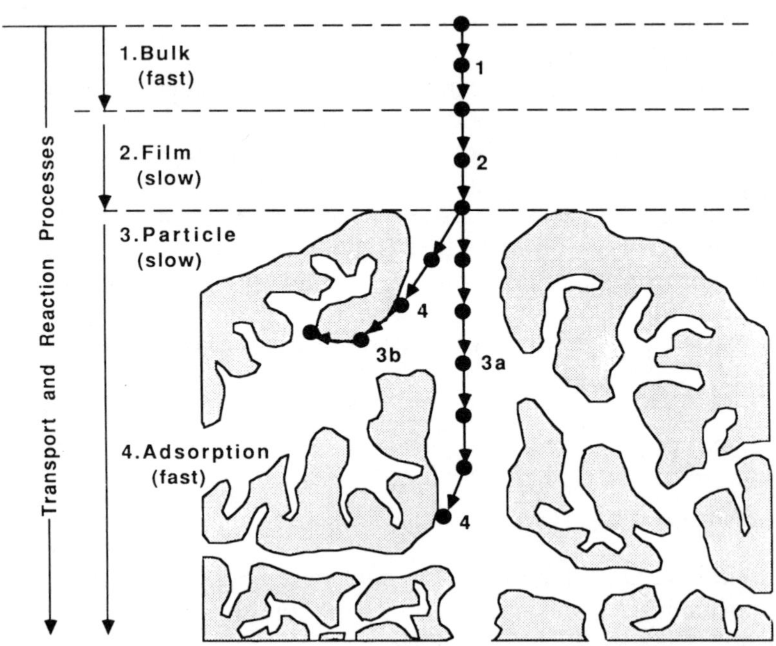

Figure 5.2. Rate-determining steps in heterogeneous soil systems. [From Weber (1984), with permission.]

the sites. The decrease in concentration of ions in these interfacial zones then is compensated by ion diffusion from the bulk solution. Thus, in most soil and soil constituent systems, either PD and/or FD is rate-limiting. Diffusion in exchangers such as soils is slower than in water because of tortuous diffusion paths (Fig. 5.2).

Rate Laws for Film and Particle Diffusion Phenomena

Rate laws for batch and flow conditions to describe FD and PD phenomena were developed by Boyd *et al.* (1947) and are given below. For PD with infinite volume (flow),

$$U(t) = 1 - \frac{6}{\pi^2} \sum_{n=1}^{\infty} \frac{1}{n^2} \exp(-\bar{D}t\pi^2 n^2/r_0^2) \tag{5.10}$$

where $U(t)$ is the fractional attainment of equilibrium at time t, n is a natural number from 0 to ∞, and r_0 is the radius of the particles.

For the batch condition (finite volume) and for a PD reaction (Sharma *et al.*, 1970),

$$U(t) = 1 - \frac{2}{3w} \sum_{n=1}^{\infty} \frac{\exp(-S_{n^2}\pi)}{1 + S_{n^2}/9w(w-1)} \tag{5.11}$$

when $w = \bar{C}_A\bar{V}_A/C_A V_A$ and $\bar{C}_A$ and C_A are the concentration of A^{z+} in the particles of volume $\bar{V}$ and in solution of volume V, respectively, $\pi = \bar{D}t/r_0^2$, and the S_n are the roots of the equation $S_n(\cot S_n) = 1 + S_n^2/3w$.

For FD control and flow conditions,

$$U(t) = 1 - \exp\left(-\frac{3D_f C_A t}{r_0 \delta \bar{C}_A}\right) \tag{5.12}$$

where δ is the thickness of the film and D_f is the film diffusion or inter-diffusion coefficient.

For FD control and for finite volume,

$$U(t) = 1 - \exp\left[\frac{3D(\bar{V}_A\bar{C}_A + V_A C_A)}{r_0 \delta \bar{C}_A V_A}\right] \tag{5.13}$$

Finally, in the intermediate range when both FD and PD control and flow conditions are present,

$$U(t) = \frac{6\theta^2}{r_0^2} \sum_{n=1}^{\infty} \frac{A_n \sin^2(m_n r_0)}{m_n^4} \exp(-\bar{D}m_n^2 t) \tag{5.14}$$

where $\theta = C_A D/\bar{C}_A \bar{D}\delta$, $A_n = m_n^2 r_0^2 + (\theta r_0 - 1)^2/m_n^2 r_0^2 + (\theta r_0 - 1)\,\theta r_0$, and m_n is the roots of the equation $m_n r_0 = (1 - \theta r_0) \tan m_n r_0$.

Differentiating between Film- and Particle-Diffusion-Controlled Phenomena

The rates of most ion exchange reactions on soils and soil constituents are FD- and/or PD-controlled. Any factor that lowers the rate of PD without causing a decrease in the FD rate will favor PD control. Any factor that increases the FD rate will also favor control by PD. Helfferich (1966) proposed the following criteria for determining whether PD or FD is rate-limiting.

$$\frac{\bar{C}\bar{D}\delta}{CD_f r_0}(5 + 2\alpha_{A/B}) \ll 1 \qquad \text{PD control}$$

$$\frac{\bar{C}\bar{D}\delta}{CD_f r_0}(5 + 2\alpha_{A/B}) \gg 1 \qquad \text{FD control}$$

where C is the total counterion concentration, δ is the film thickness, and the separation factor $\alpha_{A/B} \equiv \bar{C}^*_A C_B / \bar{C}^*_B C_A$ where C_A and C_B are the concentrations of counterions A and B and $\bar{C}^*_A$ and $\bar{C}^*_B$ are the equilibrium concentrations of counterions A and B.

Vigorous Mixing. Film diffusion is reduced or eliminated with vigorous mixing.

Flow Velocity. Mortland and Ellis (1959) found that the rate of potassium release from vermiculite was affected by flow velocity. On this basis, the authors concluded that film diffusion was a rate-controlling step. However, Reed and Scott (1962) speculated that there was little possibility as in the case of Na-tetraphenylboron extracting solutions that film diffusion was a factor as suggested by Mortland and Ellis (1959). Reed and Scott (1962) speculated that the rate-determining process in the extraction of potassium from vermiculite was particle diffusion.

Sparks *et al*. (1980b) investigated the effect of different flow velocities on the rate of potassium desorption from the B22t horizon of a Dothan soil (Fig. 5.3). At any given time, the amount of potassium desorbed increased as flow velocity increased. These data are the result of more displacing calcium passing through the column at the faster flow velocity. The potassium adsorption in this system is also concentration-dependent. High flow velocities allow the system to remain dynamic because of the low potassium concentration in solution. The apparent equilibrium between K–Ca exchange favored calcium adsorption and potassium desorption at higher flow velocities. The slower flow velocity would provide a greater contact time between solution desorbed potassium and the soil. In the

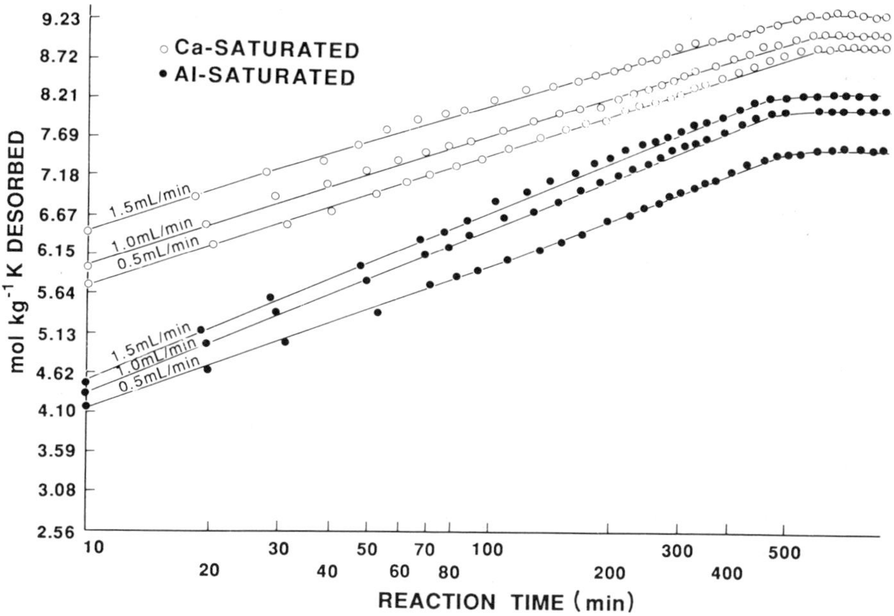

Figure 5.3. Potassium desorption at 298 K using 0.01 *M* $CaCl_2$ from soil (Al- and Ca-saturated) as a function of time plotted on a semilogarithmic scale and at flow velocities of 0.5, 1.0, and 1.5 ml min^{-1}. [From Sparks *et al.* (1980b), with permission.]

Dothan soils, the rate of potassium desorption increased slightly with flow velocity (Fig. 5.3). The k'_d values increased little with flow velocity. Sparks *et al.* (1980b) concluded from these observations that PD-controlled exchange was rate limiting for potassium desorption from the Dothan soils.

Hydrodynamic Film Thickness. The hydrodynamic condition of the system is an important factor determining whether a kinetic process is rate-limited by FD or PD. The thickness of the hydrodynamic film may be affected by stirring or shaking in a batch process, flow in a column process, the hydration status of cations, and the ionic strength of the background electrolytes. The rate of FD will be affected concurrently without affecting the rate of PD. However, a decrease in film thickness will tend to favor PD-controlled exchange (Reichenberg, 1957).

Particle Size. Particle size usually affects the type of diffusion that predominates in an ion exchange process (Mortland and Ellis, 1959; Helfferich, 1966). Film diffusion usually predominates with small particles and PD is usually rate-limiting for large particles.

Solution Concentration. A solution of low concentration usually favors FD-controlled processes, while a solution of high concentration favors PD control.

Magnitude of E Values. Film diffusion typically has E values of 17–21 kJ mol^{-1}, while PD is characterized by E values of 21–42 kJ mol^{-1} (Boyd *et al*., 1947; Reichenberg, 1957). However, just relying on the magnitude of E values to differentiate between FD and PD is tenuous.

Interruption Tests. Perhaps the most reliable method for determining whether FD or PD is rate-limiting is the use of interruption tests first employed by Kressman and Kitchener (1949). With these tests, one removes the exchanger particles from the solution for a short time during the reaction process. During interruption, concentration gradients within the particle disappear. If PD is rate-limiting, the rate immediately after reimmersion of the particles is greater than immediately before interruption. If FD is rate-limiting, interruption should have no effect on the rate. An example of data before and after interruption is shown in Fig. 5.4.

Hoinkins *et al*., (1967) glued the ion exchange particles to a cylinder that rotated in the solution and could be easily removed and reimmersed. Bunzl (1974b) attached peat particles to a polyvinylchloride (PVC) cylinder with self-adhesive tape. The cylinder, which had a small propeller at its lower end, was rotated at 470 rpm in an H solution to saturate the peat. Then the cylinder was dipped in $Pb(NO_3)_2$ and removed at certain times. Interruption tests showed that FD was rate-limiting for Pb–H exchange on peat.

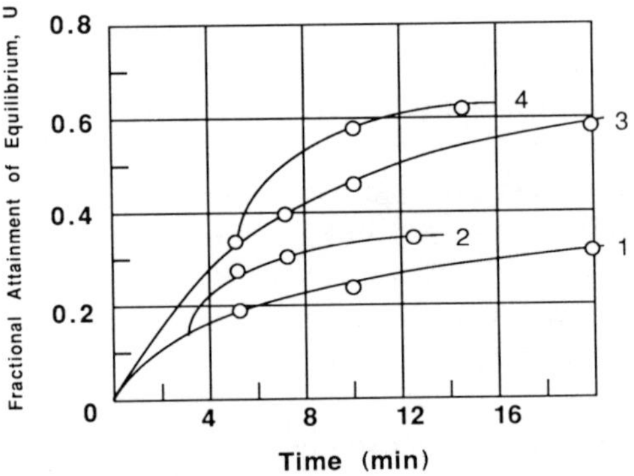

Figure 5.4. Use of interruption technique to study the rate of exchange between NH_4^+–resin and NEt_4^+ (bromide) and Th^{4+} (nitrate) at 298 K: 1, Th^{4+} (nitrate); 2, Th^{4+} (nitrate) interrupted at 2.5 min for 20 min; 3, NEt_4^+ (bromide); and 4, NEt_4^+ (bromide) interrupted at 5 min for 30 min. [From Kressman and Kitchener (1949), with permission.]

Quantification and Elucidation of Rate-Limiting Steps

Differentiating between CR and mass-transfer processes (PD and FD) is relatively easy since mass transfer is rate-limiting for most exchange processes on soils and soil constituents. Unfortunately, differentiating between FD and PD rate processes is difficult. In many cases, ion exchange is limited by both types of diffusion, particularly when the rate of ion transfer by both types is the same.

Ogwada and Sparks (1986c) presented a theory using combined observed rate coefficients under static, stirred, and vortex mixing conditions and quantified FD, PD, and CR resistances for ion exchange on a soil. This theory and its application is presented next.

A one-dimensional horizontal adsorption process was considered whereby the adsorption of ions from solution by soil particles occurs in a series rather than a parallel reaction mode, with three main consecutive steps: FD, PD, and CR. Since these processes are sequential steps, the slowest of the three is rate-controlling.

Consider a reactive species A in a solution in contact with soil particles. Its concentration at the interface where the reaction is taking place is C_A. The rate of the reaction at the surfaace will therefore depend on the concentration at the interface, which is itself directly dependent on the rate of the actual adsorption mechanism at the surface or at interlayer sites. Assuming a first-order reaction, the rate is defined as

$$v_A = k_R C_A \tag{5.15}$$

where v_A is the rate of reaction of A at the soil colloid sites and k_R is the rate coefficient for the reaction.

The depletion of C_A at the interface has to be compensated by diffusion from the bulk solution. The usual algebraic equation for this process is given in terms of an appropriate driving force (Frank-Kamenetskii, 1979; Froment and Bischoff, 1979) as

$$J_A = k_{Dif}(C_S - C_A) \tag{5.16}$$

where J_A is the mass flux with respect to the fixed particle surfaces, k_{Dif} is the mass transfer coefficient, and C_S is the concentration of A in the bulk solution.

Under steady-state conditions, the rate of diffusion of A from the bulk solution to the interface is approximately equal to the rate of adsorption of A, and this can be used to eliminate the unmeasured interface concentration C_A. Thus,

$$v_A \simeq J_A = v_{AS} \tag{5.17}$$

where v_{AS} is the rate of reaction of A under the steady-state conditions.

Combining Eq. (5.15) and (5.16) gives

$$C_A = k_{Dif} C_S/(k_R + k_{Dif}) \tag{5.18}$$

and

$$v = [k_{Dif} k_R/(k_R + k_{Dif})]\, C_S = k_O\, C_S \tag{5.19}$$

where

$$k_O = k_{Dif} k_R/(k_R + k_{Dif}) \tag{5.20}$$

This relation can assume a simple form. Instead of the reaction-rate coefficient (k_R), mass-transfer coefficient (k_{Dif}), and experimentally observed rate coefficient (k_O), one can use the reciprocals of the coefficients such that

$$1/k_O = 1/k_R + 1/k_{Dif} \tag{5.21}$$

One now has both kinetic and diffusional resistances that are additive. However, ion adsorption processes in soils or clay minerals are assumed to occur in a series rather than in a parallel-reaction mode. As noted earlier, the diffusion steps include both FD and PD processes. One can therefore assign two resistances to the diffusion resistance,

$$1/k_{Dif} = 1/k_P + 1/k_F \tag{5.22}$$

where k_P is the PD rate coefficient and k_F is the FD rate coefficient. Equations (5.21) and (5.22) can be combined to give

$$1/k_O = 1/k_R + 1/k_P + 1/k_F \tag{5.23}$$

The greatest of these resistances (smallest of the rate coefficients) will be the rate-determining step under static (no agitation) adsorption conditions.

The observed diffusion and reaction rate coefficients can be obtained from specific experiments. To quantify the rate coefficients on the right-hand side of Eq. (5.23), kinetic experiments could be conducted such that the global rate is preferably determined by FD, PD, or CR. In the laboratory these steps can be simulated separately by conducting experiments using static, stirred, or vortex batch adsorption systems (Ogwada and Sparks, 1986b). Therefore, to these systems one can assign additive resistance relations as follows:

Static system $\quad 1/k_{os} \simeq 1/k_R + 1/k_P + 1/k_F \quad$ (5.24)

Stirred system $\quad 1/k_{ot} \simeq 1/k_R + 1/k_P \quad$ (5.25)

Vortex batch system $\quad 1/k_{ov} \simeq 1/k_R \quad$ (5.26) where k_{os}, k_{ot}, and k_{ov} are observed rate coefficients in the static, stirred, and vortex batch adsorption systems, respectively.

The rate coefficients obtained experimentally from these three systems represent the full spectrum of the rate-limiting steps in ion adsorption processes. Algebraic combination of these observed rate coefficients from each of the systems can be used to quantify and isolate the rate-controlling step in the static system. These suggestions were based on the assumption that the static system has all of the three resistances. Such an assumption seems plausible based on previous work (Ogwada and Sparks, 1986b). Due to agitation in the stirred system, the influence of the FD resistance that was prevalent in the static system is greatly reduced by uniform mixing and rapid mass transfer across the hydrodynamic film (Ogwada and Sparks, 1986a,b). Therefore, the stirred system is assigned only the reaction and particle resistance relations. Ogwada and Sparks (1986c) assumed that in using the vortex batch technique, mass transfer is maximized and is at its highest rate, which limits the formation of a concentration gradient in the film or hydrated interlayer spaces of the particles. Thus, in the vortex batch system it was assumed that the observed rate coefficient approximates that of the reaction step in the static system, and they therefore assigned only one resistance relation in this system. This seems justified based on previous studies (Ogwada and Sparks, 1986a,b).

Accordingly, the parameters for Eq. (5.24) can be calculated as follows:

$$1/k_F \simeq 1/k_{os} - (1/k_R + 1/k_P) \tag{5.27}$$

or

$$1/k_F = 1/k_{os} - 1/k_{ot} \tag{5.28}$$

and

$$1/k_P = 1/k_{ot} - 1/k_R \tag{5.29}$$

or

$$1/k_P = 1/k_{ot} - 1/k_{ov} \tag{5.30}$$

As mentioned above, $1/k_{os}$, $1/k_{ot}$, and $1/k_{ov}$ are parameters obtained from respective experiments, and therefore $1/k_F$ and $1/k_P$ are the only unknowns to be calculated from Eq. (5.28) and (5.30).

In summary, Ogwada and Sparks (1986c) developed a model and assumed that the adsorption of ions from solution by soil particles occurs in a series rather than a parallel reaction mode. Thus, mass-transfer processes and CR occur consecutively. Under the steady-state approximation, the rate of mass transfer is approximately equal to the rate of the reaction, so that instantaneous change in the concentration of C_A with time approaches

TABLE 5.1 Rate Coefficients for Film Diffusion, Particle Diffusion, and Chemical Reaction Rate Processes of K^+ Adsorption under Static Conditions[a]

Adsorbent	Temperature (K)	Rate coefficients (min^{-1}) k_F	k_P	k_R
Kaolinite	283	0.031	15.060	1.444
	298	0.034	15.361	3.482
	313	0.037	15.739	4.981
Chester loam	283	0.029	0.247	0.837
	298	0.030	0.280	2.331
	313	0.035	0.322	3.240
Vermiculite	283	0.037	0.055	0.421
	298	0.042	0.060	0.945
	313	0.044	0.072	1.647

[a] From Ogwada and Sparks (1986c), with permission.

zero. Under such conditions, only some of the processes or reaction that constitute the overall mechanism are at equilibrium. The overall reaction need not be at equilibrium.

Based on this theory, Ogwada and Sparks (1986c) found that the rate-controlling process for potassium exchange on kaolinite and a soil under static conditions was FD (Table 5.1), as evidenced by the k_F values being lower than the k_P or k_R values. For vermiculite, there was no significant difference between k_P and k_F values, which suggested that both FD and PD were rate-limiting.

Chemical Reaction and Diffusion

There are some cases where a "reaction," that is, the formation or dissolution of a chemical bond, is involved along with ion exchange phenomena (Helfferich, 1983). Examples of this are acid–base neutralization, dissociation of weak electrolytes in solution or weak ionogenic groups in ion exchangers, complex formation, or combinations of these (Table 5.2). With some of these, very low apparent $\bar{D}$ in ion exchangers have been noted.

In ion exchange phenomena accompanied by reaction the increase or decrease in concentration of a species results from both mass transfer and reaction,

$$\partial C_i/\partial t = -\text{div } J_i + v_i \tag{5.31}$$

TABLE 5.2 Examples of Ion Exchange with Reaction[a]

$\boxed{RSO_3^-} + H^+$	$+ Na^+ + OH^-$	$\rightarrow$	$\boxed{RSO_3 + Na^+}$	$+H_2O$
$\boxed{RCOO^- + Na^+}$	$+ H^+ Cl^-$	$\rightarrow$	$\boxed{RCOOH}$	$+ Na^+ + Cl^-$
$\boxed{RNH_2}$	$+ H^+ + Cl^-$	$\rightarrow$	$\boxed{RNH_3^+ + Cl^-}$	
$\boxed{2RSO_3^- + Ni^{2+}}$	$+ 4Na^+ + EDTA^{4-}$	$\rightarrow$	$\boxed{2RSO_3^- + 2Na^+}$	$+ 2Na^+ + NiEDTA^{2-}$

[a] Boxes symbolize ion exchanger particles. From Helfferich (1983), with permission.

rather than from mass transfer reaction alone (Fick's second law, without reaction term) as in normal ion exchange. Helfferich (1965) found that ion exchange accompanied by reaction can cause the rate of the reaction to decrease dramatically and solution concentration to be greatly changed.

There are two ways in which a chemical reaction can affect ion exchange rates (Helfferich, 1983). One possibility is that the reaction is slow compared with diffusion. Thus, in the limit, diffusion is fast enough to cause a leveling out of any concentration gradients within the ion exchanger particle. Thus, the reaction is the sole rate-controlling factor, and rate is independent of particle size.

The second case is where the reaction is faster than diffusion but is binding ions on whose diffusion ion exchange depends. This binding inhibits the diffusion of the ions and lowers the rate of exchange (Schwarz *et al.*, 1964). The rate is thus controlled by slow diffusion, which is affected by the equilibrium of the fast reaction. Since the process is diffusion-controlled, the exchange rate is dependent on particle size. This type of ion exchange can be referred to as reaction-retarded diffusion and is much more likely to happen than is reaction control. In fact, Helfferich (1983) notes that no case of genuine reaction control has been definitively shown.

RATES OF ION EXCHANGE ON SOILS AND SOIL CONSTITUENTS

The rates of ion exchange on soils range from a few seconds to days, depending on a number of factors. These include type and quantity of inorganic and organic constituents, ion charge and radius, and kinetic methodology (Chapter 3).

Mineralogical Composition

Perhaps one of the most important factors affecting the rate of ion exchange on soils and soil constituents is clay mineralogical composition. The type of clay mineral has a significant effect on the rates of exchange. Vermiculite, smectite, kaolinite, and hydrous micas vary considerably in their ionic preferences, in ion binding affinities, and in types of ion exchange reactions. Such fundamental differences in these clay minerals account for the varying kinetics of exchange.

Rates of ion exchange on kaolinite, smectite, and illite are usually quite rapid. Sawhney (1966) found that sorption of cesium on illite and smectite was rapid, while on vermiculite, sorption had not reached an equilibrium even after 500 h (Fig. 5.5). Sparks and Jardine (1984) found that potassium adsorption rates on kaolinite and montmorillonite were rapid, with an apparent equilibrium being reached in 40 and 120 min, respectively. However, the rate of potassium adsorption on vermiculite was very slow. Malcom and Kennedy (1969) studied Ba–K exchange rates on kaolinite, illite, and montmorillonite using a potassium ion-specific electrode to monitor the kinetics. They found >75% of the exchange occurred in 3 s, which represented the response time of the electrode. The rate of Ba–K exchange on vermiculite was characterized by a rapid and slow rate of exchange.

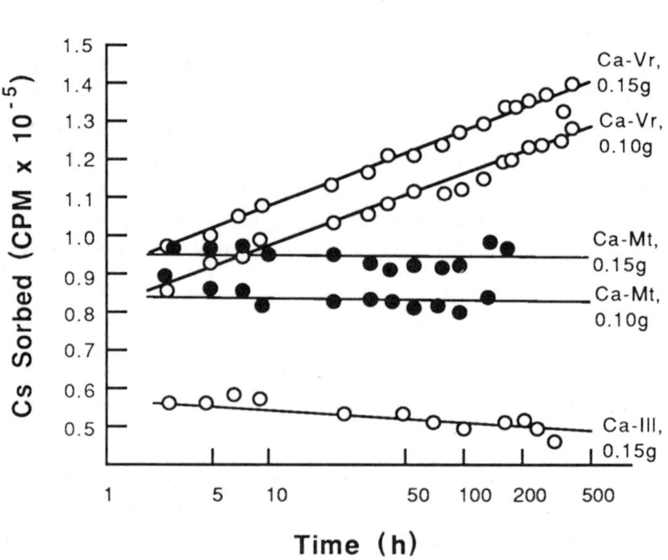

Figure 5.5. Adsorption of cesium (Cs) by Ca-saturated clay minerals with time. Ill, illite; Mt, montmorillonite; Vr, vermiculite. [From Sawhney (1966), with permission.]

The above results are related to the structural properties of the clay minerals. In the case of kaolinite, the tetrahedral layers of adjacent clay sheets are held tightly by hydrogen bonds. Therefore, only readily available planar external surface sites exist for exchange. With smectite, the inner peripheral space is not held together by hydrogen bonds, but instead it is able to swell with adequate hydration and thus allow for rapid passage of ions into the interlayer.

Rates of exchange on vermiculite and micaceous minerals, particularly involving ions such as K^+, NH_4^+, and Cs^+, are usually quite slow. These are 2 : 1 clay minerals with peripheral spaces that impede many ion exchange reactions. Micaceous minerals typically have a more restrictive interlayer space than vermiculite, since the area between layer silicates of micas is selective for certain types of cations such as K^+, Cs^+, NH_4^+, and H_3O^+ (Sparks and Huang, 1985; Sparks, 1987a).

Unlike kaolinite and montmorillonite, there are several sites for ion exchange reactions to occur on mica and vermiculite (Bolt *et al.*, 1963; Sparks and Jardine, 1984). Bolt *et al.* (1963) studied potassium exchange on mica and proposed three sites for reactivity. Slow reactions were ascribed to interlattice exchange sites, rapid reactions to external planar sites, and intermediate reactions to readily exposed edge sites. Sawhney (1966) found two distinct reaction rates for cesium exchange on a Ca-vermiculite. The first reaction was ascribed to a rapid exchange of cesium with cations on external planar surface sites and interlattice edges, followed by a second, slow reaction in which cesium diffuses into the interlayers.

The particle size fractions of a soil or sediment may also differ in their exchange rates. Kennedy and Brown (1965) measured Ca–Na exchange on sand-sized sediments and found 90% exchange occurred on 0.12–0.25 mm and 0.25–0.50 mm sand fractions in 3 and 7 min, respectively. Malcom and Kennedy (1970) studied Ba–K exchange on clay, silt, sand, and gravel size fractions of a river sediment using a potassium ion-specific electrode. Barium–potassium exchange on fine and coarse clay and fine silt was most rapid, with >75% exchange occurring within 3 s and complete exchange in 2 min. The exchange rates on medium and coarse silts and very fine sand diminished with increasing particle size. Only 30–50% exchange occurred within 3 s, and complete exchange required 5–10 min. With fine and medium sands, <20% exchange occurred within 3 s, and complete exchange required 20 min to 1 h. Malcom and Kennedy (1970) found that the exchange rate of coarse sand, fine gravel, and medium gravel was so slow that practically no exchange was observed within 3 s and <50% exchange occurred in 1 h. Complete exchange required 1–2 days. The low rates of exchange on the coarse fractions were ascribed to slow diffusion of ions through the weathering rind on these fractions.

Few data are available on the rate of ion exchange on organic matter and humic substances, but one should not ignore organic constituents in kinetic investigations. Bunzl and co-workers (Bunzl, 1974a,b; Bunzl *et al.*, 1976) have studied kinetics of ion exchange on soil organic matter. Generally, the rates of exchange are rapid. For example, the $t_{1/2}$ for adsorption and desorption of Pb^{2+}, Cu^{2+}, Cd^{2+}, Zn^{2+}, and Ca^{2+} on peat ranged from 5 to 15 s.

Ion Charge and Radius

The charge of an ion has a significant affect on diffusion rate through an ion exchanger such as a resin (Helfferich, 1962a). Generally, the rate of exchange decreases as the charge of the exchanging species increases. Sharma *et al.*, (1970) studied the exchange rates of Cs^+, Co^{2+}, and La^{3+} in H^+, Ca^{2+}, and La^{3+}. The $\bar{D}$ values calculated for the trace ions are shown in Table 5.3 and reveal that the rate of exchange decreases rapidly as the ionic charge increases, that is, $\bar{D}_{Cs^+} > \bar{D}_{Co^{2+}} > \bar{D}_{La^{3+}}$. Moreover, it can be seen that the $\bar{D}$ values for a particular ion such as Cs^+ decrease as the charge of the other ion increases, that is, $\bar{D}_{Cs^+-H^+} = 625 \times 10^{-8}$, $\bar{D}_{Cs^+-Ca^{2+}} = 120 \times 10^{-8}$, $\bar{D}_{Cs^+-La^{3+}} = 27.4 \times 10^{-8}$ cm^2 s^{-1}.

The ionic radius can also affect the rate of ion exchange. An example of this is shown in the work of Sharma *et al.*, (1970), who studied the rate of exchange of La^{3+}, Tb^{3+}, and Lu^{3+} in HCl using a flow system (Fig. 5.6). Calculated $\bar{D}$ values are shown in Table 5.4, along with data on ion size. Lanthanum has the largest ionic crystal radius, while Lu^{3+} has the smallest. The hydrated size of the ions is in the order $Lu^{3+} > Tb^{3+} > La^{3+}$.

TABLE 5.3 Calculated Particle Diffusion Coefficients[a] $\bar{D} \times 10^8$ cm^2 s^{-1}

	Microcomponent		
Macrocomponent	Cs^+	Co^{2+}	La^{3+}
H^+ (1.0 *M*)	625	94	10.4
Ca^{2+} (2.0 *M*)	120	13	4.3[b]
La^{3+} (0.5 *M*)	27	3.5	1.6

[a] From Sharma *et al.* (1970), with permission.

[b] This experiment was done using the exchange of Tb^{3+} with Ca^{2+}, and $\bar{D}_{Tb^{3+}}$ was found to be 8. Since $\bar{D}_{Tb^{3+}}/\bar{D}_{La^{3+}}$ was found to be about 2, a value of 4 was estimated for $\bar{D}_{La^{3+}}$ in Ca^{2+}.

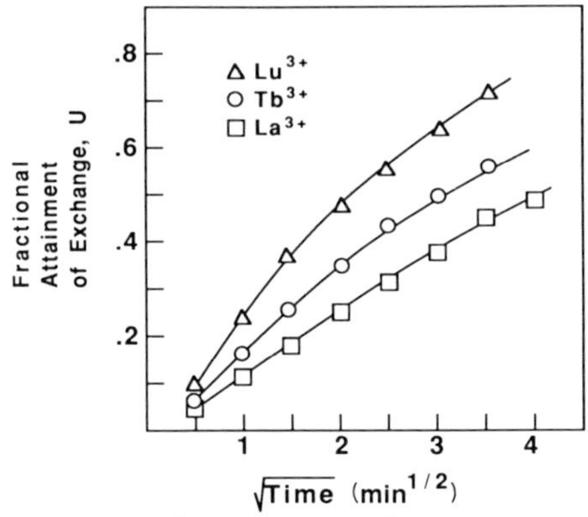

Figure 5.6. Rate of exchange of La^{3+}, Tb^{3+}, and Lu^{3+} in 1.94 *M* hydrochloric acid in a flow system. [From Sharma *et al.* (1970), with permission.]

TABLE 5.4 Calculated Particle Diffusion Coefficients $\bar{D}$ and Other Pertinent Data for La^{3+}, Tb^{3+}, and Lu^{3+} Ions[a]

	La^{3+}	Tb^{3+}	Lu^{3+}
$\bar{D} \times 10^8$ cm^2 s^{-1} in 1.94 *M* HCl	8.7	16.3	35.0
Ionic (crystal) radius (nm)	0.106	0.092	0.085

[a] From Sharma *et al.* (1970), with permission.

BINARY CATION AND ANION EXCHANGE KINETICS

Exchange on Soils and Inorganic Soil Constituents

All of the research on ion exchange kinetics on soils and inorganic soil constituents until the present has primarily involved binary processes. Studies have been conducted with numerous cations and anions that are important from both plant nutrition and environmental viewpoints. Details of some of these studies are given in Table 5.5. A variety of kinetic models has been used to describe these various adsorptive–adsorbent reactions. It is also clear from each of these studies that the type of ion and adsorbent greatly affects the rates of exchange, as was pointed out earlier. Additionally, the results of these studies are also affected by the kinetic method used. There is currently a great need to study ternary and quaternary kinetics of exchange on soils, since there are seldom situations under field conditions where only two ions are participating in exchange reactions.

TABLE 5.5 Published Studies on Binary Cation and Anion Exchange Kinetics on Soils and Inorganic Soil Constituents

Cation	Exchanger	References
Cation exchange kinetics		
Aluminum	Clay minerals	Jardine *et al.* (1985b)
Ammonium	Soils	Carski and Sparks (1987)
Cadmium	Soils	Cavallaro and McBride (1978), Amacher *et al.* (1986)
Cesium	Clay minerals	Sawhney (1966), Komareni (1978), Noll *et al.* (1986)
Chromium	Soils	Amacher *et al.* (1986)
Copper	Soils	Harter and Lehmann (1983), Jopony and Young (1987)
Lead	River muds	Salim and Cooksey (1980)
Mercury	Soils	Amacher *et al.* (1986)
Nickel	Soils	Harter and Lehmann (1983)
Potassium	Clay minerals	Keay and Wild (1961), Malcom and Kennedy (1969), Sparks and Jardine (1984), Yan and Xue (1987), Burns and Barber (1961)
	Soils	Talibudeen and Dey (1968), Sivasubramaniam and Talibudeen (1972), Selim *et al.* (1976a), Feigenbaum and Levy (1977), Sparks *et al.* (1980a,b), Sparks and Jardine (1981), Sparks and Rechcigl (1982), Jardine and Sparks (1984), Ogwada and Sparks (1986a,b,c), Sharpley (1987)
Anion exchange kinetics		
Arsenite	Soils	Elkhatib *et al.* (1984a,b)
Borate	Soils	Griffin and Burau (1974), Evans and Sparks (1983), Peryea *et al.* (1985)
Chloride	Soils	Thomas (1963), Pasricha *et al.* (1987)
Nitrate	Soils	Pasricha *et al.* (1987)
Phosphate	Clay minerals	Atkinson *et al.* (1970), Kuo and Lotse (1972)
	Calcite	Griffin and Jurinak (1974)
	Resins	Amer *et al.* (1955), Evans and Jurinak (1976)
	Sediments	Li *et al.* (1972)
	Soils	Dalal (1974), Barrow and Shaw (1975), Novak and Adriano (1975), Enfield *et al.* (1976), Evans and Jurinak (1976), Fiskell *et al.* (1979), Vig *et al.* (1979), Chien and Clayton (1980), Van Riemsdijk and de Hann (1981), Sharpley *et al.* (1981a,b), Onken and Matheson (1982), Lin *et al.* (1983), Sharpley (1983), Polyzopoulos *et al.* (1987)
Silica	Oxides	Hingston and Raupach (1967)
	Soils	Brown and Mahler (1987)
Sulfate	Oxides	Hackerman and Stephens (1954), Rajan (1978), Tripathi *et al.* (1975)
	Soils	Chang and Thomas (1961), Hodges and Johnson (1987), Pasricha *et al.* (1987)

Exchange on Humic Substances

Most of the studies involving ion exchange kinetics on soil constituents have been concerned with inorganic components such as clay minerals, oxides, etc., as just discussed. Rates of ion exchange on humic substances, while extremely important, have not been extensively studied.

Bunzl and co-workers have investigated the kinetics of ion exchange of several metals, including Ca^{2+}, Cd^{2+}, Cu^{2+}, Pb^{2+}, and Zn^{2+}, on humic acid and peat (Bunzl, 1974a,b; Bunzl *et al.*, 1976). In each of these studies, the rate-limiting step for ion exchange of a metal ion for H_3O^+ was FD.

Figure 5.7(a), taken from Bunzl *et al.* (1976), shows the sorption and desorption of Cu^{2+} and Ca^{2+} on H-saturated peat as a function of time. For sorption, 0.05 mmol of Cu^{2+} or Ca^{2+} was added to a suspension containing 0.1 g H-saturated peat in 0.2 l water. For desorption, an equivalent amount of H_3O^+ ions was added to the partially Cu^{2+}- or Ca^{2+}-converted peat sample in 0.2 l water. Within an experimental error of $\pm$ 6%, the sorption of 1 mol of metal cation charge was accompanied by the release of 1 mol of H_3O^+ charge. Bunzl *et al.* (1976) concluded that this equivalence was indicative not only of an ion exchange mechanism but also of the formation of a metal chelate, involving displacement of H^+ from the acidic groups of a humic substance. Figure 5.7(b) shows that proton exchange on peat is rapid and the rate of metal sorption on H-peat is higher than the rate of metal desorption.

Figure 5.8 (Bunzl *et al.*, 1976) shows the initial rates of sorption and desorption during the first 10 s of exchange and corresponding half times for Pb^{2+}, Cu^{2+}, Cd^{2+}, Zn^{2+}, and Ca^{2+} by H-saturated peat using the same concentrations of metal and H_3O^+ added for the experiments shown in Fig. 5.7. The absolute initial rates of sorption decreased in the order Pb > Cu > Cd > Zn > Ca, which is the order observed for the calculated distribution coefficients. This indicates that the higher the selectivity of peat for a given metal ion, the faster the initial rate of sorption. The relative rates of sorption, as shown by half-times (Fig. 5.8), shows that Ca^{2+} was sorbed the fastest, followed by $Zn^{2+} > Cd^{2+} > Pb^{2+} > Cu^{2+}$. Thus, even though the absolute rate of Ca^{2+} adsorption by peat was low, the relative rate was comparatively high, since the total amount of Ca^{2+} adsorbed was small. The relative rates of desorption, as illustrated by the half-times, show longer times for Pb^{2+}, Cu^{2+}, and Ca^{2+} but shorter ones for Cd^{2+} and Zn^{2+}.

There has been great concern about how heavy metals affect environmental quality. Heavy metals (As, Cu, Cd, Cr, Pb, Hg, and Ag) come from a number of different sources, including industrial sources, domestic water supplies, residential wastewater, surface runoff, atmospheric

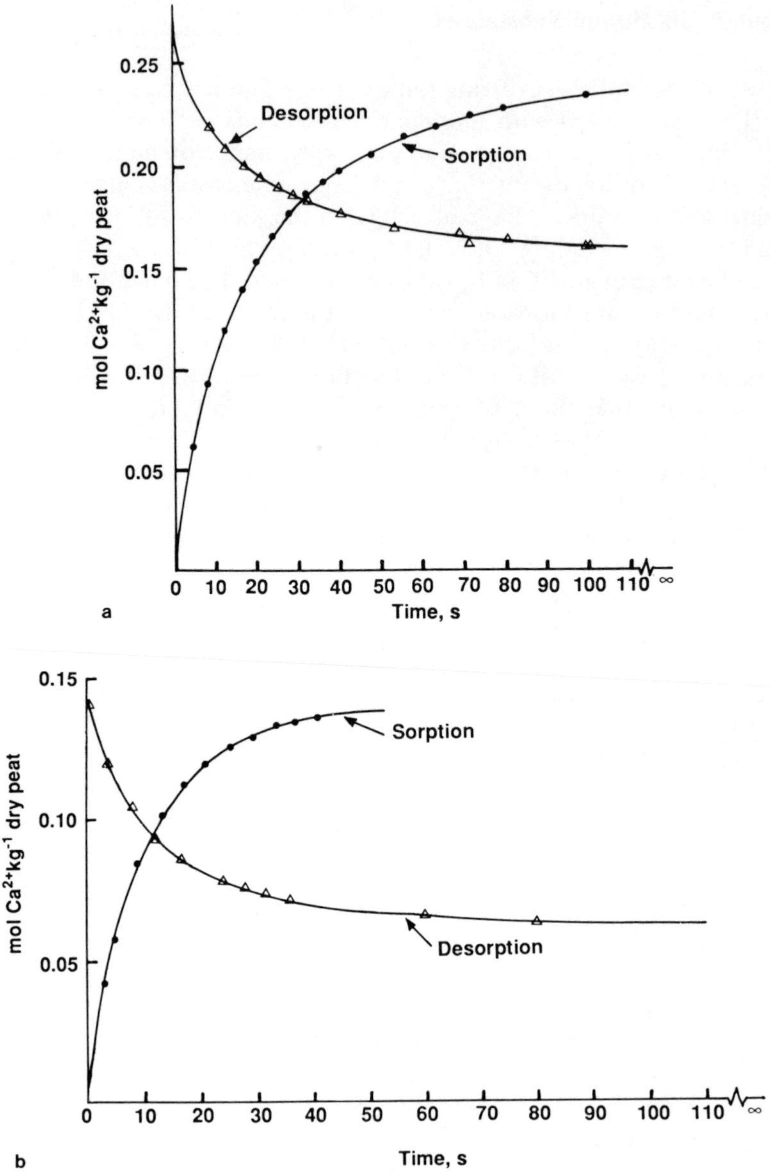

Figure 5.7. Amount of (a) copper (Cu^{2+}) and (b) calcium (Ca^{2+}) sorbed and subsequently desorbed by peat as a function of time. Sorption involved addition of 0.2 cmol Cu^{2+} or Ca^{2+} kg^{-1} H-saturated peat in 1 liter of water. Desorption involved addition of 0.2 cmol H_3O^+ ions to the samples from the sorption experiments in 1 liter water. The stirring rate was 470 rpm and the temperature of the studies was 298 K. [From Bunzl *et al.* (1976), with permission.]

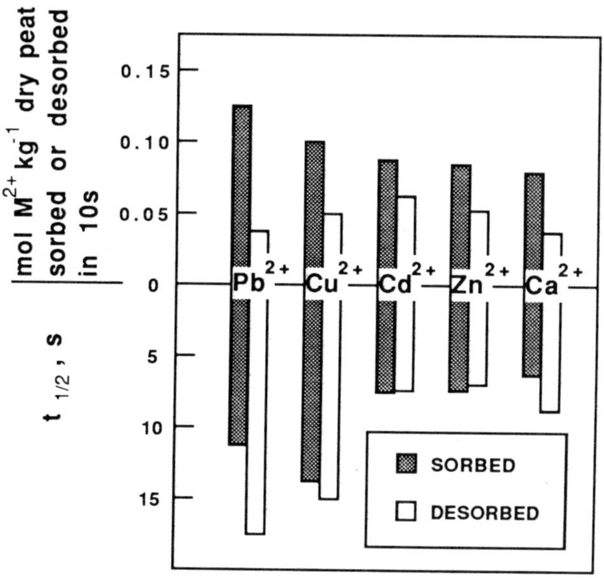

Figure 5.8. Initial rates of sorption and subsequent desorption during the first 10 s of exchange on peat and corresponding half-times of lead (Pb^{2+}), copper (Cu^{2+}), cadmium (Cd^{2+}), zinc (Zn^{2+}), and calcium (Ca^{2+}) obtained from similar curves and experimental conditions as given in Fig. 5.7. [From Bunzl *et al.* (1976), with permission.

precipitation, groundwater flow, and infiltration. One of the major processes for removing heavy metals is bioadsorption. Activated sludge solids are composed mainly of a mixture bacteria, protozoa, and fungi. Thus, to mitigate the deleterious effects of heavy metals on our environment, it is imperative to better understand the rates and mechanisms of heavy metal interactions with sludges. However, few such studies have been conducted, and this is an area in need of more research.

The kinetics of metal sorption by sludges is rapid (Neufeld and Heiman, 1975; Nelson *et al.*, 1981). Neufeld and Heiman (1975) concluded initially that metal sorption on sludges is not a biological phenomenon but is related to the physical and chemical properties of the sludge. Furthermore, they found that metal uptake was not affected by heavy metal concentration and organism viability.

Tien (1987) studied the kinetics of heavy metal sorption–desorption on sludge using the stirred-flow reactor method of Carski and Sparks (1985). Sorption–desorption reactions were rapid with an equilibrium reached in ~30 min. The sorption–desorption reactions were reversible. The sorption rate coefficients were of the order Hg > Pb > Cd > Cu > Zn > Co > Ni, while the desorption rate coefficients were of the order Cd > Cu > Hg >

Ni > Zn > Pb > Co. Tien (1987) noted that the preference of sludge solids for the sorption of one metal ion over another appeared to be affected by the covalent radius. Some studies have shown that the bonding between heavy metals and functional groups on the surface of microorganisms is probably covalent. For example, Crist *et al.* (1981) suggested that the covalent bonding of amino and carboxyl groups is responsible for the adsorption of Cu(II) ions in the cell walls of the alga *Vaucheria*.

TERNARY ION EXCHANGE KINETICS

Few studies have appeared on ternary exchange kinetics, even though in most soil systems exchange processes involve three or more ions. Bajpai *et al.* (1974) studied Mn–Cs–Na, Ba–Mn–Na, and Sr–Mn–Cs exchange on a resin. Nernst–Planck equations were used to study the ternary systems, and when FD effects were included in the equations they described the data well. Other ternary kinetic studies were conducted by Wolfrum *et al.* (1983) and Plicka *et al.* (1984).

Plicka *et al.* (1984) studied Mg–UO_2–Na exchange on a resin using the kinetic model described below. If one assumes that ions A and B are sorbed by a resin in the C form and that the exchange reactions begin at time $t = 0$, then

$$z_c A^{z_a} + z_a \bar{C}^{z_c} \rightleftharpoons z_c \bar{A}^{z_a} + z_a C^{z_c} \tag{5.31}$$

$$z_c B^{z_b} + z_b \bar{C}^{z_c} \rightleftharpoons z_c \bar{B}^{z_b} + z_b C^{z_c} \tag{5.32}$$

where the bars indicate the exchange phase.

This is a three-component (A, B, C) system, and the transport or diffusion of each component can be described using the Nernst–Planck equation. The relation of the diffusion flux of the i component ($i, j, k \equiv$ A, B, C) can be obtained by combining three Nernst–Planck equations, for the respective components, under the conditions of zero electric current inside the resin particle:

$$\begin{aligned} \bar{J}_i = -[&\bar{D}_i\bar{D}_j z_j(z_j\bar{C}_j\,\nabla\bar{C}_i - z_i\bar{C}_i\,\nabla\bar{C}_j) \\ &+\bar{D}_i\bar{D}_k z_k(z_k\bar{C}_k\,\nabla\bar{C}_i - z_i\bar{C}_i\,\nabla\bar{C}_k)]/Z \end{aligned} \tag{5.33}$$

and

$$Z = \sum_{i=1}^{3} z_i^2 \bar{C}_1 \bar{D}_1$$

A complicated second-order differential equation is obtained by substituting Eq. (5.33) into the equation of continuity and by transformation into polar coordinates (for the spherical particles of the ion exchanger):

$$\frac{\partial \bar{C}_i}{\partial t} = \Bigg\{\bar{D}_i\bar{D}_j z_j\left[z_j\left(\frac{\partial \bar{C}_i}{\partial r_d}Z - \bar{C}_j\frac{\partial Z}{\partial r_d}\right)\frac{\partial \bar{C}_i}{\partial r_d} - z_i\left(\frac{\partial \bar{C}_i}{\partial r_d}Z - \bar{C}_i\frac{\partial Z}{\partial r_d}\right)\frac{\partial \bar{C}_j}{\partial r_d}\right]$$
$$+\bar{D}_i\bar{D}_k \bar{z}_k\left[z_k\left(\frac{\partial \bar{C}_k}{\partial r_d}Z - \bar{C}_k\frac{\partial Z}{\partial r_d}\right)\frac{\partial \bar{C}_i}{\partial r_d}\right.$$
$$\left.\left. - z_i\left(\frac{\partial \bar{C}_i}{\partial r_d}Z - \bar{C}_i\frac{\partial Z}{\partial r_d}\right)\frac{\partial \bar{C}_k}{\partial r_d}\right]\right\}\Big/Z^2 + \Bigg\{\bar{D}_i\bar{D}_j z_j\left[z_j\bar{C}_j\left(\frac{\partial^2 \bar{C}_i}{\partial r_d^2} + \frac{2}{r_d}\frac{\partial \bar{C}_i}{\partial r_d}\right)\right.$$
$$\left. - z_i\bar{C}_i\left(\frac{\partial^2 \bar{C}_j}{\partial r_d^2} + \frac{2}{r_d}\frac{\partial \bar{C}_j}{\partial r_d}\right)\right] + \bar{D}_i\bar{D}_k z_k\left[z_k\bar{C}_k\left(\frac{\partial^2 \bar{C}_i}{\partial r_d^2} + \frac{2}{r_d}\frac{\partial \bar{C}_i}{\partial r_d}\right)\right.$$
$$\left.\left. - z_i\bar{C}_i\left(\frac{\partial^2 \bar{C}_k}{\partial r_d^2} + \frac{2}{r_d}\frac{\partial \bar{C}_k}{\partial r_d}\right)\right]\right\}\Big/Z \tag{5.34}$$

The time concentration profiles $\bar{C}_i(t, r_d)$ of the components (A, B, C) in the particle of the resin were found by the solution of the system consisting of three respective equations [Eq. (5.34)]. The total quantity of each component in the particle, which is related to the volume of the particle in time t, can be evaluated by substitution of the $\bar{C}_i(t, r_d)$ values into Eq. (5.35):

$$q_i(t) = \frac{3}{r_0^3}\int_0^{r_0} \bar{C}_i(t, r_d) r_d^2 \, dr_d \tag{5.35}$$

In Eqs. (5.31)–(5.35) for i, j, k (= A, B, C), $\bar{C}_i$ is the molar concentration in the resin, mol m^{-3}; $\bar{J}_i$ is the diffusion flux in the resin particle, mol m^{-2} s^{-1}; $\bar{D}_i$ is the diffusion coefficient of the ion in the resin, m^2 s^{-1}; q_i is the total mass quantity of the i component in the particle, related to its volume, mol m^{-3}; r_o is the particle radius in meters, which is assumed to be equal for all particles; r_d is the distance from the particle center, $r_d t(0, r_0)$ in meters; and t is time in seconds.

DETERMINATION OF THERMODYNAMIC PARAMETERS FROM ION EXCHANGE KINETICS

The connection between chemical equilibria and completely reversible reactions has been known for a long time (e.g., see Glasstone *et al.*, 1941; Laidler, 1965). Keay and Wild (1961) determined enthalpy and entropy values from Na–Mg exchange kinetic studies on vermiculite. Sparks and Jardine (1981) calculated apparent thermodynamic parameters for K–Ca exchange phenomena on soil from apparent adsorption and desorption rate coefficients, k'_a and k'_d, respectively. The magnitude of the apparent thermodynamic parameters compared favorably with pseudothermodynamic parameters calculated from Eyring's reaction rate theories. One shortcoming of the previous two studies was that the thermodynamic parameters calculated from kinetic measurements were not compared to those determined from classical exchange isotherm data. As Ogwada and Sparks (1986a) showed, to properly determine equilibrium parameters from rate data, a number of assumptions and experimental measurements must first be made.

The general cation exchange reaction in a binary exchange system can be expressed as (Sposito, 1986):

$$z_2M_1\,X_{z_1}\,(\text{s}) + z_1M_2^{z_2}(\text{aq}) = z_1M_2X_{z_2}(\text{s}) + z_2M_1^{z_1}(\text{aq}) \qquad (5.36)$$

where the cations $M_i(i = 1,2)$ have valences $z_i(i = 1,2)$. In Eq. (5.36), 1 mol of an exchangeable cation M_i reacts with z_i mol of exchanger charge.

The rate of this reaction can be given as (Denbigh, 1981; Sposito, 1986) as

$$d\xi/dt = v_1 - v_{-1} \qquad (5.37)$$

where $d\xi$ = extent of the reaction parameter, v_1 is the rate of the forward, and v_{-1} is the rate of the backward reaction in Eq. (5.36). One can denote z_iv_1 as the rate of which M_2X_{z2} in Eq. (5.36) is formed and z_iv_{-1} is the rate at which it is consumed, that is, the same reaction describes both the forward and backward processes in cation exchange.

The individual rates v_1 and v_{-1} are affected by temperature, pressure, and the concentrations of the species in Eq. (5.36). At equilibrium, the left side of Eq. (5.37) will disappear and $v_{1\text{eq}}/v_{-1\text{eq}}$ where eq is the equilibrium condition will be a function of temperature, pressure, and the equilibrium composition of the exchanger and aqueous solution phase. Because the activities of the species in Eq. (5.36) have an identical dependence, $v_{1\text{eq}}/v_{-1\text{eq}}$ depends on temperature, pressure, and the species activities (Denbigh, 1981). But this same relationship applies to the quotient of the right and left sides of Eq. (5.38) for the determination of the exchange equilibrium constant (K_{ex}) for the reaction in Eq. (5.36), which can be expressed as,

$$K_{\text{ex}} = \frac{(M_2X_{z_2})^{z_1}(M_1^{z_1})^{z_2}}{(M_1X_{z_1})^{z_2}(M_2^{z_2})^{z_1}} \qquad (5.38)$$

where the terms in parentheses denote activity. Thus,

$$v_{1eq}/v_{-1eq} = F[(M_2X_{z_2})_{eq}^{z_1} (M_1^{z_1})_{eq}^{z_2}/(M_1X_{z_1})_{eq}^{z_2} (M_2^{z_2})^{z_1} K_{ex}] \quad (5.39)$$

Equation (5.39) gives a general relationship between reaction rate and thermodynamics, predicated on Eq. (5.37) and the uniqueness of Eq. (5.36) as experimental facts of kinetics. If it further assumed that $F(x) = x$ on the right side of Eq. (5.39), then V_{1eq}/V_{1eq} will equal the quotient of the middle and left side of Eq. (5.38):

$$V_{1eq}/V_{-1eq} = (M_2X_{z_2})_{eq}^{z_1} (M_1^{z_1})_{eq}^{z_2}/(M_1X_{z_1})_{eq}^{z_2}(M_2^{z_2})_{eq}^{z_1} K_{ex} \quad (5.40)$$

Equation (5.40) is expected whenever v_1 and v_{-1} depend on powers of the concentrations of the reactants and products in Eq. (5.36), and the power exponents are the stoichiometric coefficients of the four species involved (Denbigh, 1981). Even if Eq. (5.37) is simplified in this way, the assumption must be made that the mechanism of the cation exchange reaction at equilibrium does not change when $v_1 \neq v_{-1}$ so that finite rate data can be used to apply Eq. (5.40).

If the following exchange reaction is studied kinetically,

$$Ca_{1/2}\,X(s) + K^+(aq) = KX(s) + \tfrac{1}{2}\,Ca^{2+}(aq) \quad (5.41)$$

it can be shown (Ogwada and Sparks, 1986a) that Eqs. (5.37)–(5.40) are applicable *only* if diffusion is not the rate-controlling step for the reaction in Eq. (5.41). If the rate-controlling process in Eq. (5.41) is diffusion, then information about K_{ex} cannot be derived from an analysis of kinetics data (Ogwada and Sparks, 1986a). If diffusion can be eliminated or practically so, the rates v_1 and v_{-1} can be modeled for K-Ca exchange using the equations

$$v_1 = k_a m_s q_{ca} m_K \qquad v_{-1} = k_d m_s q_K m_{Ca}^{1/2} \quad (5.42)$$

where m_s is the mass of the exchanger, q_i (i = Ca or K) is the number of moles of charge of metal i adsorbed by unit mass of the exchanger, and m_i is a molality (i = Ca or K). The rate of formation of KX(s) now can be given by Eq. (5.43):

$$d\xi/dt = m_s(dq_K/dt) = m_s[k_a q_{Ca} m_K - k_d q_K m_{Ca}^{1/2}] \quad (5.43)$$

Once k_a and k_d values are determined through standard kinetic analysis (Sparks, 1986), one can determine the relationship $K_{ex} = k_a/k_d$. Then $\Delta G^\circ_{ex} = -RT \ln K_{ex}$, where ΔG°_{ex} is the standard free energy. Using the Arrenhius and van't Hoff equations [Eqs. (2.55) and Eqs. (2.59) and (2.60), respectively], E_a and E_d values can be determined [Eq. 2.60)]. From the van't Hoff equation, the standard enthalpy of exchange (ΔH°_{ex})

TABLE 5.6 Comparison of Equilibrium and Kinetic Approaches for Determining Thermodynamics of Potassium Exchange in Soils[a]

	Equilibrium approach				Kinetic approach, continuous flow			
Temperature (K)	K_{eq}	ΔG° (kJ mol^{-1})	ΔH° (kJ mol^{-1})	ΔS° (J mol^{-1} K^{-1})	K_{eq}	ΔG° (kJ mol^{-1})	ΔH° (kJ mol^{-1})	ΔS° (J mol^{-1} K^{-1})
Chester loam								
283	9.59	−5.32		−38.72	1.55	−1.03		−9.14
298	6.77	−4.74		−38.69	1.38	−0.80		−9.58
308	4.24	−3.70		−40.83	1.33	−0.74		−9.44
			−16.28				−4.45	
Downer sandy loam								
283	9.81	−5.37		−50.72	2.16	−1.81		−12.09
298	6.43	−4.61		−50.72	1.97	−1.68		−12.22
308	5.27	−4.26		−50.21	1.82	−1.53		−12.15
			−19.73				−5.05	

	Kinetic approach, batch				Kinetic approach, vigorously mixed batch			
Temperature (K)	K_{eq}	ΔG° (kJ mol^{-1})	ΔH° (kJ mol^{-1})	ΔS° (J mol^{-1} K^{-1})	K_{eq}	ΔG° (kJ mol^{-1})	ΔH° (kJ mol^{-1})	ΔS° (J mol^{-1} K^{-1})
Chester loam								
283	2.93	−2.53		−16.80	5.45	−3.99		−16.19
298	2.62	−2.39		−16.52	4.66	−3.82		−15.96
308	2.27	−2.10		−16.83	4.01	−3.57		−16.24
			−7.72				−9.14	
Downer sandy loam								
283	2.49	−2.15		−36.73	11.55	−5.76		−41.81
298	2.76	−2.52		−33.64	7.53	−5.00		−42.24
308	1.48	−1.00		−37.45	6.19	−4.67		−42.00
			−15.44				−18.53	

[a] From Ogwada and Sparks (1986c), with permission.

can be determined:

$$d \ln K_{ex}/dT = \Delta H^\circ_{ex}/RT^2 \tag{5.44}$$

or

$$E_a = E_d = \Delta H^\circ \tag{5.45}$$

Then, ΔS°_{ex}, the standard entropy of exchange, can be determined from

$$\Delta S_{ex} = \frac{\Delta H^\circ_{ex} - \Delta G^\circ_{ex}}{T} \tag{5.46}$$

Ogwada and Sparks (1986a) found that thermodynamic parameters calculated from exchange isotherms and from the kinetics approach outlined above compared well in trend (Table 5.6) and gave the same inferences of ion behavior for two soils. However, the data clearly show that except for the vigorously mixed batch technique, where mass-transfer phenomena were significantly reduced, the magnitude of the thermodynamic parameters for the two approaches compared poorly. The ΔG° values for both soils at all three temperatures calculated using the vigorously mixed batch technique compared well to those calculated using the equilibrium approach. However, with batch and continuous flow techniques, where pronounced diffusion-controlled exchange occurs, the comparison was poor.

These data support the earlier contention that if the rate-controlling process is diffusion and not the reaction in Eq. (5.41), then *no* information about equilibrium can be derived from kinetic analyses. If a kinetics experiment can be designed so that diffusion is significantly reduced, then one can use a kinetics approach to gather thermodynamic information about soils or other heterogeneous systems.

SUPPLEMENTARY READING

Bunzl, K., Schmidt, W., and Sansoni, B. (1976). Kinetics of ion exchange in soil organic matter. IV. Adsorption and desorption of Pb^{2+}, Cu^{2+}, Cd^{2+}, Zn^{2+}, and Ca^{2+} by peat. *J. Soil Sci.* **27,** 32–41.

Helfferich, F. (1962). "Ion Exchange." McGraw-Hill, New York.

Helfferich, F. (1966). Ion exchange kinetics. *In* "Ion Exchange" (J. A. Marinsky, ed.), Vol. 1, pp. 65–100. Dekker, New York.

Helfferich, F. (1983). Ion exchange kinetics—Evolution of a theory. *In* "Mass Transfer and Kinetics of Ion Exchange" (L. Liberti and F. Helfferich, eds.), pp. 157–179. Matinus Nijhoff Publishers, Dordrecht, The Netherlands.

Jackman, A. P., and Ng, K. T. (1986). The kinetics of ion exchange on natural sediments. *Water Resour. Res.* **22,** 1664–1674.

Liberti, L., and Passino, R. (1985). Ion exchange kinetics in selective systems. *In* "Ion Exchange and Solvent Extraction" (J .A. Marinsky and Y. Marcus, eds.), pp. 175–210. Dekker, New York.

6

Kinetics of Pesticide and Organic Pollutant Reactions

INTRODUCTION

The fate of pesticides and organic pollutants in natural waters and in soils is strongly dependent on their sorptive behavior (Karickhoff, 1980). Sorption affects not only physical transport of these materials but also their degradation. It is also important to note that the chemical reactivity of pollutants in a sorbed state may be different from their behavior in aqueous solution. Karickhoff (1980) notes that sorbents such as inorganic and organic soil constituents may affect solution-phase processes by changing the solution-phase pollutant concentration or by affecting the release of pollutants into the solution phase.

The sorptive behavior of pesticides and organic pollutants can be studied from either equilibrium or kinetic viewpoints. While both are important, perhaps the time-dependent processes are least understood. As environmental concerns intensify about groundwater pollution, waste disposal, and soil detoxification, it will become increasingly important to better understand the kinetics and mechanisms of pesticide and organic pollutant interactions with soils. For comprehensive treatments on pesticides and organic pollutants in soils the reader is encouraged to consult books and reviews by Weber and Gould (1966), Kearny and Helling (1969), Helling *et al.* (1971), Goring and Hamaker (1972), Weber (1972), Guenzi (1974), Hance (1981), Rao and Davidson (1982), and Saltzman and Yaron (1986).

PESTICIDE SORPTION AND DESORPTION KINETICS

Classes of Pesticides

The primary soil components responsible for pesticide sorption are clay minerals and, especially, humic materials. Pesticides can be divided into cationic, basic, acidic, and nonionic classes (Saltzman and Yaron, 1986).

The predominant sorption mechanism of cationic pesticides such as $diquat^{2+}$ and $paraquat^{2+}$ on soils is apparently ion exchange (Best *et al.*, 1972). An ion exchange mechanism was also observed for sorption of cationic pesticides on clays by Weber *et al.* (1965). Other sorption mechanisms for this class of pesticides could be hydrogen bonding, ion-dipole, and physical forces. Mortland (1970) has shown that cationic pesticide sorption on clays is significantly affected by the pesticide's molecular weight, functional groups, and molecular configuration.

Basic herbicides include the *s*-triazines. They can be sorbed on clays by cationic sorption. However, this mechanism is affected by the acidity of the medium or, most significantly, the clay mineral's surface acidity (Bailey and White, 1970). When the surface acidity of the clay is greater than two pH units higher than the dissociation constant of the sorptive, sorption occurs mainly by van der Waals forces (Saltzman and Yaron, 1986).

As with other pesticides, organic matter appears to be the soil constituent most important for basic pesticide sorption. Weber *et al.* (1969) showed that maximum sorption of several *s*-triazines on organic matter occurred at pH levels close to the pK_a values of the compounds. The molecular structure of the pesticide and the pH of the sorbent strongly affected the degree of sorption. The pH-dependent sorption and the relationship between pH and dissocation constant with pH suggests an ion exchange mechanism (Saltzman and Yaron, 1986).

Other mechanisms for sorption of cationic pesticides include hydrogen bonding and coordination between sorbate and the exchangeable cation. When the surface acidity of the clay is one or two pH units lower than the dissociation constant, chemical sorption could occur.

Senesi and Testini (1982) have studied sorption of *s*-triazines on humic acids using elemental, thermal, infrared, and spin resonance analyses.

Using infrared analysis they showed that one of the binding mechanisms is ionic bond formation, following proton transfer from the humic acids to the *s*-triazine molecules. Another mechanism found by Senesi and Testini (1982) was hydrogen bonding.

Acidic pesticides such as 2,4-D, 2,4,5-T, picloram, and dinoseb can ionize in aqueous solutions forming anionic species (Saltzman and Yaron, 1986). Sorption of these pesticides on soils has also been correlated with soil organic matter content (Hamaker *et al.*, 1966), and in their anionic form they can be sorbed on soils, clays, and amorphous materials at low pH. The mechanisms of sorption for these compounds are proton association and, for the molecular form, van der Waals sorption (Saltzman and Yaron, 1986). Hydrogen bonding and electrostatic interactions are other possible mechanisms for sorption.

As Saltzman and Yaron (1986) have noted, most of the pesticides currently used are nonionic. These include compounds belonging to such chemically different groups as chlorinated hydrocarbons, organophosphates, carbamates, ureas, anilines, anilides, amides, uracils, and benzonitriles. They differ widely in their sorptive behavior on soils, but studies conducted thus far again point to the importance of organic matter. Neutral organic materials are only slightly sorbed by clay minerals. But at low water contents, significant sorption could occur. The major sorption mechanisms are formation of cation-dipole and coordination bonds, hydrophobic and hydrogen bonds, and van der Waals attraction.

Reaction Rates

The rate of sorption and desorption of pesticides on soils and soil constituents has been investigated by a number of workers (see, e.g., Hance, 1967) and is dependent on the type of sorbent, pesticide, and rate of mixing. For example, sorption seems much slower on humic substances (Khan, 1973). Other factors that may affect the kinetics are swelling of the sorbent and temperature (Hance, 1967).

Hance (1967) investigated the rate of sorption and desorption of four pesticides (monuron, linuron, atrazine, and chlorpropham) on two soils, a soil organic matter fraction, and bentonite, a 2 : 1 smectitic clay mineral. An equilibrium in sorption was reached in 24 h for every system except one (Table 6.1). With eight of the 18 systems equilibrium was reached in less than 4 h, and in five cases equilibrium was established in 1 hr. Equilibrium was attained for most of the systems in 4–24 h. Desorption was slower than sorption. In only eight systems was an equilibrium reached in 24 h. Hance

TABLE 6.1 Summary of the Periods Required for the Establishment of Sorption and Desorption Equilibria of Pesticides on Soils and Soil Constituents[a]

Time for establishment of equilibrium (h)	Chemical	Adsorbent	
		Sorption	Desorption
0–4	Monuron	Bentonite	—
	Linuron	Nylon, silica gel	Nylon, silica gel
	Atrazine	Dark gray loam, sandy loam, bentonite	Dark gray loam, bentonite
	Chlorpropham	Dark gray loam, organic matter	Dark gray loam
4–24	Monuron	Dark gray loam, sandy loam	—
	Linuron	Dark gray loam, sandy loam, organic matter, bentonite	Organic matter
	Atrazine	Organic matter	—
	Chlorpropham	Sandy loam, bentonite	Sandy loam, organic matter
24–72	Monuron	Organic matter	Organic matter
	Linuron	—	Bentonite
	Atrazine	—	Sandy loam
> 72	Monuron	—	Dark gray loam, sandy loam, bentonite
	Linuron	—	Dark gray loam, sandy loam
	Atrazine	—	Organic matter
	Chlorpropham	—	Bentonite

[a] From Hance (1967), with permission.

(1967) found desorption equilibrium was attained in <4 h for five pesticide–colloid systems, while 72 h was not sufficient time for seven systems to attain an equilibrium. Reversibility existed for almost all the systems.

The sorption and desorption of pesticides by soils and soil constituents such as clay minerals and humic substances has generally been characterized by an initial rapid rate folllowed by a much slower approach to an apparent equilibrium (Haque *et al.*, 1968; Leenheer and Ahlrichs, 1971; Khan, 1973; McCall and Agin, 1985). The initial reaction(s) have been associated with diffusion of the pesticides to and from the surface of the sorbent, while the slower reaction(s) have been related to PD of the pesticides into and out of micropores of the sorbent.

An example of this kind of relationship is shown in the work of Leenheer and Ahlrichs (1971) for sorption of carbaryl and parathion on soil humic materials (Fig. 6.1). An apparent equilibrium was attained in 2 h or less,

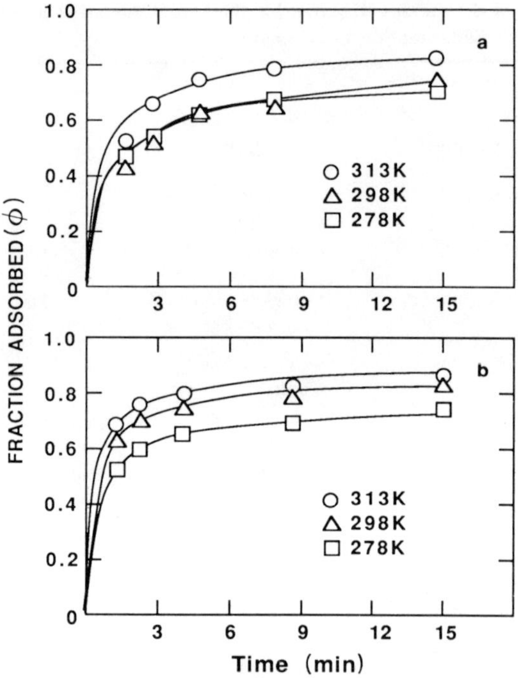

Figure 6.1. Rate of approach to equilibrium of (a) carbaryl and (b) parathion on Ca-saturated soil humic material. [From Leenheer and Ahlrichs (1971), with permission.]

and the amount of pesticide sorbed at equilibrium decreased as temperature increased. Haque *et al.* (1968) observed a slow rate of sorption of 2,4-D on illite, montmorillonite, and kaolinite (Fig. 6.2).

Steinberg *et al.* (1987) studied the persistence of 1,2-dibromoethane (EDB) in soils and found that low amounts of the organic were released with time, particularly if EDB had not been freshly added to the soil (Fig. 6.3). They suggested that the slow release rate was due to EDB being trapped in soil micropores where release is influenced by extreme tortuosity and/or steric restrictions. It was estimated that based on a radial diffusion model, 23 and 31 years would be required for a 50% equilibrium in EDB release to occur from two Connecticut soils. The previous studies point out that while sorption of pesticides is usually rapid and often reversible in the laboratory, extraction from field soils is extremely slow and often requires multiple extractions or even chemical dissolution of the soil matrix.

The parabolic diffusion law (Chapter 2) has successfully described rate data for 2,4,5,-T and parathion (Weber and Gould, 1966) as well as other

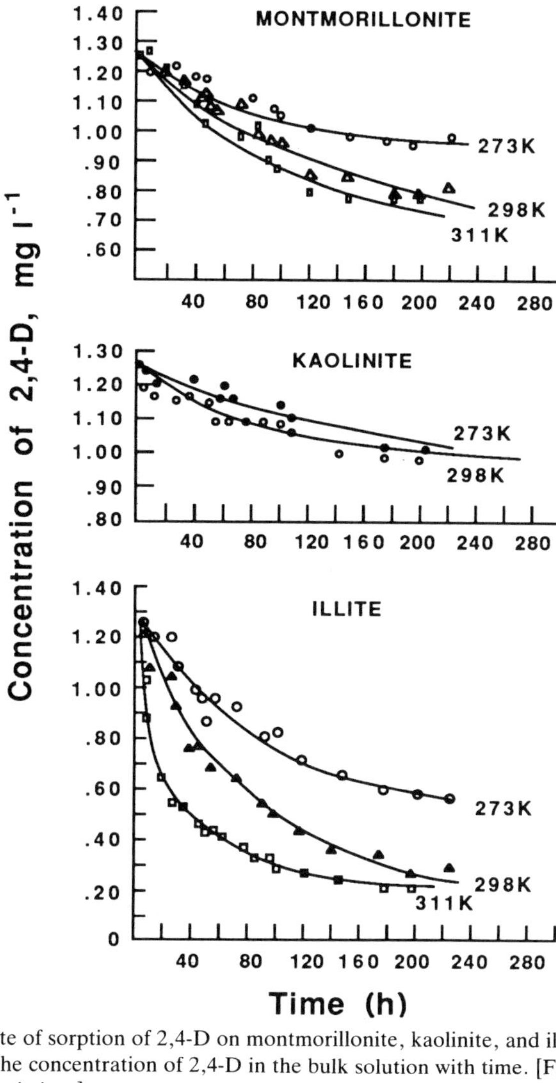

Figure 6.2. Rate of sorption of 2,4-D on montmorillonite, kaolinite, and illite, as shown by the decrease in the concentration of 2,4-D in the bulk solution with time. [From Haque *et al.* (1968), with permission.]

pesticides on carbon, indicating that PD is the rate-limiting step. A relative sorption rate coefficient can be calculated from the slope of a linear parabolic diffusion plot. Some representative values for sorption of pesticides on carbon are reported in Table 6.2. The rate coefficients for the different pesticides are similar, which suggests that similar rates of removal from solution may occur (Weber and Gould, 1966).

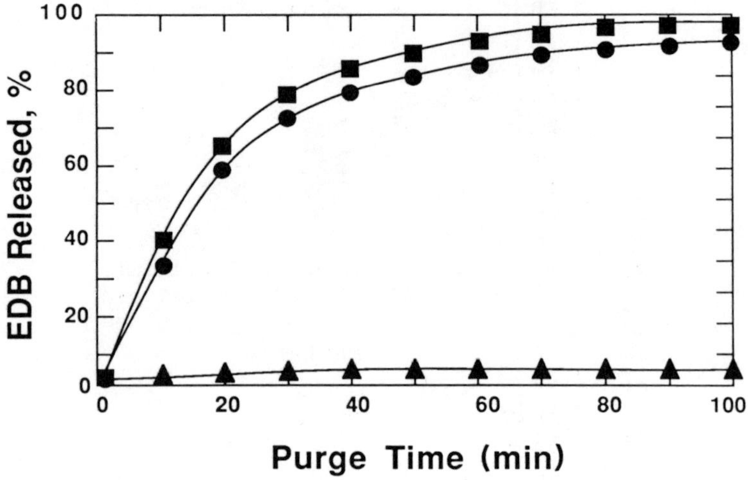

Figure 6.3. Removal of EDB from soil suspension by N_2 purging: removal of native EDB from soil (bottom curve), removal of a [^{14}C]EDB spike (10 ng g^{-1}) from the same soil after a 3-h equilibration period (middle curve), removal of [^{14}C]EDB (10 ng ml^{-1}) from distilled water. [From Steinberg *et al.* (1987), with permission.]

TABLE 6.2 Relative Sorption Rate Coefficients for Pesticide–Carbon Interactions[a]

Compound	Relative rate coefficient k (μmol g^{-2} h^{-1} × 10^{-4})
2,4-Dichlorophenoxyacetic acid	1.44
2,4,5-Trichlorophenoxyacetic acid	1.00
2-(2,4,5-Trichlorophenoxy) propionic acid	0.71
2,4-Dinitro-*o*-*sec*-butylphenol	1.35
2,4-Dinitro-*o*-*sec*-cyclohexylphenol	1.12
1-Naphthyl-*N*-methylcarbamate	1.64
O,*O*-Diethyl-*o*-*p*-nitrophenyl phosphorothioate	1.49
4-Nitrophenol	Anion, 0.52
	Neutral, 1.21
2,4-Dinitrophenol	0.87
2,4-Dinitro-*o*-cresol	1.12
2,4-Dinitrothymol	0.67
2,6-Dinitro-*p*-cyclohexylphenol	0.90

[a] From Weber and Gould (1966), with permission.

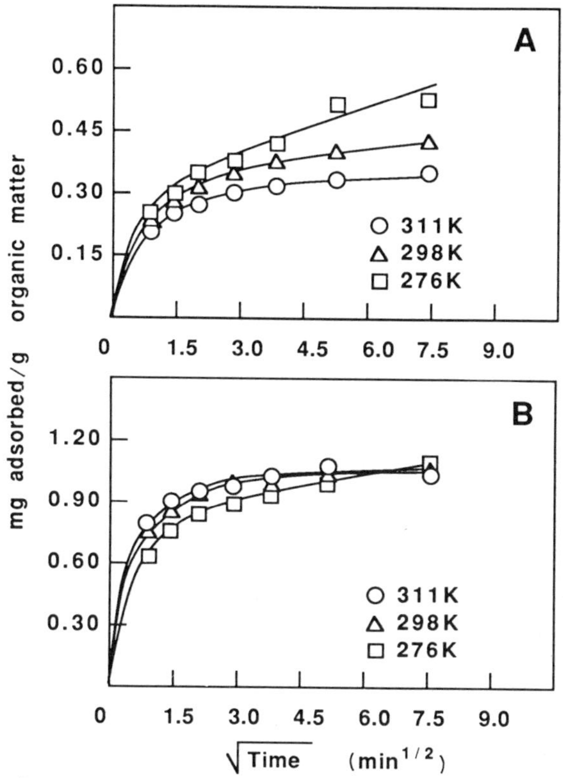

Figure 6.4. Sorption rate curves of (a) carbaryl and (b) parathion on Ca-saturated soil humic material as a function of $(\text{time})^{1/2}$. [From Leenheer and Ahlrichs (1971), with permission.]

However, when data from many of the kinetics studies on pesticide–soil interactions were plotted according to the parabolic diffusion equation, initial nonlinearity resulted (Fig. 6.4). This suggested that only at longer times did the reaction process conform to PD. The rate-limiting step for this reaction is diffusion into or out of micropores.

To determine the rate-limiting step for the initial time periods, the generalized equilibrium theory of Fava and Eyring (1956) has been employed (Haque and Sexton, 1968; Haque *et al.*, 1968; Leenheer and Ahlrichs, 1971). Equation (6.1) below takes into account both sorption and desorption in determining a rate equation, which is actually what takes place in nonflow systems. Also, in deriving Eq. (6.1) it is assumed that the reverse reaction or the desorption rate is small enough to be neglected. This assumption is satisfied by using large amounts of sorbent

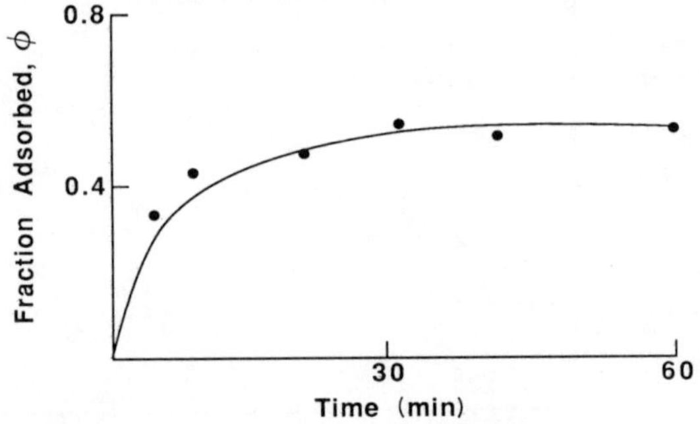

Figure 6.5. Use of generalized equilibrium theory to study rate of sorption of 2,4-D on humic acid at 278 K. [From Khan (1973), with permission.]

and low sorptive concentrations. Equation (6.1) can be expressed as

$$\frac{dU}{dt} = 2k(1 - U)\ \sinh b(1 - U) \tag{6.1}$$

where U is the distance from equilibrium as a fraction of the initial distance from equilibrium or the fraction sorbed, b is a constant, and k is the rate coefficient for sorption when $U = 0$. A plot of U versus t is shown in Fig. 6.5. The differential dU/dt for various values of U for the initial times of sorption can then be measured as the slope of the tangent to the plot at certain values of U and t. Leenheer and Ahlrichs (1971) then used a computer program that solved Eq. (6.1) for the rate coefficient k for a series of substituted b values raised in increments of 0.1 from 1 to 10. The variance of k for six known values of U/dt and $(1-U)$ was determined for each substituted k value. The solution of the rate equation was obtained when the ratio of the variance of k over the mean of k reaches a minimum for a certain value of b using Eq. (6.1).

Table 6.3 lists k and E values for the initial reaction between a number of pesticides and soil components. The k values would indicate a rapid reaction rate, and the E values are rather low, suggesting diffusion-controlled exchange. Thus, it would appear that the initial sorption of pesticides on soil components involves physisorption and that the rate depends on pesticide diffusion through the water film surrounding the sorbent, or FD.

Nonsingularity of Pesticide–Soil Interactions

Many investigators have found that in soil–pesticide studies sorption–desorption phenomena may be nonsingular or even irreversible (Farmer

TABLE 6.3 Sorption Rate Coefficients (*k*) and Energies of Activation (*E*) for Initial Reaction between Pesticides and Soil Components

Soil component	Pesticide	$k(s^{-1})$ at 298 K	E ($kJ\ mol^{-1}$)	Reference
Illite	2,4-D	8.8×10^{-7}	14.67	Haque *et al.* (1968)
Kaolinite	2,4-D	21.9×10^{-7}	13.51	Haque *et al.* (1968)
Smectite	2,4-D	5.5×10^{-7}	23.05	Haque *et al.* (1968)
Humic acid[a]	2,4-D	2.1×10^{-4}	6.70	Haque and Sexton (1968)
Soil humic acid	2,4-D	6.5×10^{-5}	8.38	Khan (1973)
Soil humic acid	Carbaryl	2.0×10^{-4}	24.72	Leenheer and Ahlrichs (1971)
Soil humic acid	Parathion	1.9×10^{-4}	26.40	Leenheer and Ahlrichs (1971)
Soil humic acid	Picloram	10.1×10^{-5}	7.54	Khan (1973)

[a] Synthetically prepared in the laboratory.

and Aochi, 1974; Rao, 1974; van Genuchten *et al.*, 1974; Murray *et al.*, 1975; Peck, 1977). Nonsingularity occurs when, for a given equilibrium solution concentration, more pesticide is held on the soil during the desorption phase than during the sorption phase. In such systems, the amount of pesticide can be significantly overestimated if predictions are made on only the sorption isotherm parameters (Bowman and Sans, 1985). Mukhtar (1976) found the degree of nonsingularity was directly proportional to the amount of pesticide sorbed prior to the start of desorption, the rate of sorption, and the sorption energy.

Rao and Davidson (1980) give three major causes for nonsingularity: (1) artifacts created due to the kinetic method, (2) failure to obtain complete equilibrium during the sorption phase, and (3) chemical and/or microbial transformation of the pesticide during the experiment. However, "true" hysteresis can occur, particularly when significant amounts of humic substances are present (Bowman and Sans, 1985). In these systems, hysteresis may slowly increase as equilibration time increases. Peck *et al.* (1980) found that as organic matter content of the sorbent increased, nonsingularity or hysteresis was more pronounced for sorption–desorption of diuron. This observation could be due to a secondary slow sorption process on organic matter (Bowman and Sans, 1985).

Most soil–pesticide sorption–desorption studies have used batch techniques, which create several problems. In many batch studies the slow portion of the soil–pesticide interactions may not be seen if observation times are too short (McCall and Agin, 1985). Additionally, desorption is usually begun by centrifuging the equilibrated soil–pesticide system, removing a known volume of pesticide solution, replacing with the same volume of pesticide-free solution, and resuspending the soil–pesticide solution. This procedure is then repeated to develop desorption isotherms initiated from a particular point on the sorption isotherm. Then there is repeated centrifugation and resuspension of the soil followed by prolonged agitation (Rao and Davidson, 1980). As noted earlier (Chapter 3), long periods of agitation can cause abrasion of the soil or colloidal particles, increasing surface area. Consequently, the number of sorption sites could be increased during the desorption phase of the experiment.

Rao *et al.* (1979) investigated whether the centrifugation–resuspension step in the batch method could be causing nonsingularity. They measured sorption–desorption isotherms for several soil–pesticide systems using the batch method given above and two modifications of this method. These methods eliminated centrifugation and were (1) a water-immiscible organic solvent as a third phase to desorb the pesticide from the soil and aqueous phases (three-phase method), and (2) desorption by dilution of the soil–water–pesticide system (dilution method). With the batch method, nonsingularity was always observed. The modified batch methods gave identical sorption–desorption isotherms for some pesticide–soil systems and nonsingular results for others. Rao *et al.* (1979) thus concluded that the pesticide sorption mechanism was important in determining whether or not centrifugation caused nonsingularity. They recommended that the dilution method be used, but could not explain why centrifugation could cause nonsingularity. However, Horzempa and DiToro (1983) used a variation of the dilution method and found that centrifugation did not cause nonsingularity in their experiments.

Bowman and Sans (1985) studied the degree of desorption hysteresis with pesticides and soils and clays using a consecutive (centrifugation–resuspension) method and a dilution method (Fig. 6.6), and their results agree with those of Rao *et al.* (1979). Hysteresis was minimal with the dilution method but was more pronounced when the consecutive desorption method was used (except for a fensulfothion sulfone–Ca-illite system). Bowman and Sans (1985) hypothesized that maybe the sorbent was compacted during centrifugation, increasing the time required for desorption processes to reestablish an equilibrium. This could cause one to conclude that partitioning had shifted in favor of the sorbent. However, there is evidence that with sorption repeated centrifugation does not affect partitioning as with desorption (Bowman, 1979).

Not reaching an equilibrium between the soil and pesticide before desorption is begun could also cause nonsingularity. Diffusion of pesticides into soil micropore sites associated with clay minerals and organic matter could cause a pseudo-equilibrium (Hance, 1967; Rao *et al.*, 1979).

Chemical and microbiological transformations of the pesticide while the experiment is being conducted could also cause nonsingularity in pesticide–soil sorption–desorption isotherms. Hamaker and Thompson (1972) theorized that a portion of a sorbed pesticide is more strongly held than

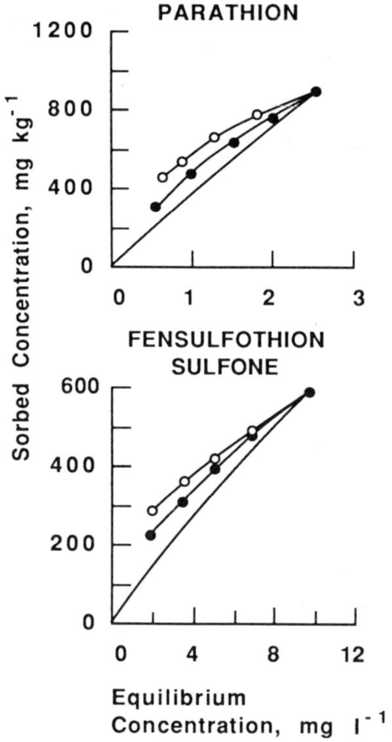

Figure 6.6. Sorption (solid lines) and desorption isotherms where ○ represents consecutive desorption method and ● indicates dilution desorption method for parathion and fensulfothion sulfone in aqueous suspensions of an organic soil showing hysteresis, [From Bowman and Sans (1985), with permission.]

the remainder and that this fraction tends to increase with time. Microbial degradation of the pesticide during an experiment could also cause nonsingularity.

DEGRADATION RATES OF PESTICIDES

Another aspect of pesticide–soil interactions that is very important in predicting the effect of pesticides on environmental quality is the degradation rate. It is not the purpose of this discussion to give an in-depth discussion of pesticide degradation in soils. Numerous reviews are available on transformations, metabolic pathways, persistence, and $t_{1/2}$ values of pesticides and other toxic organics in soils (Hamaker, 1966, 1972; Crosby, 1973; Kaufman, 1976; Laveglia and Dahm, 1977), and the reader is encouraged to read these sources for thorough discussions. Rather, the purpose of this section of the chapter is to briefly discuss degradation kinetics of pesticides in soils.

Numerous studies have shown that several factors affect pesticide degradation rates, including soil type, water content, pH, temperature, and clay and organic matter content (Rao and Davidson, 1980). Hamaker (1972) has published an excellent review on the quantitative aspects of pesticide degradation rates in soils. He consider two types of rate models:

(1) $$dC/dt = -kC^n \tag{6.2}$$

(2) $$dC/dt = -v_{max}[C/(x + C)] \tag{6.3}$$

where C is concentration (mg l^{-1}), t is time (days), k is the rate constant ($days^{-1}$), n is the reaction order, v_{max} is the maximum rate, and x is a constant. When $n = 1$, Eq. (6.2) reduces to a first-order kinetic equation. Walker (1976a,b) found that degradation of simazine, prometryne, and linuron herbicides in field soils conformed to first-order kinetics.

Table 6.4 shows first-order rate coefficients and $t_{1/2}$ values for degradation of a number of pesticides in soils (Rao and Davidson, 1982). The k and $t_{1/2}$ values calculated from field data are based on the disappearance of the parent compound (solvent extractable). Table 6.4 also includes k and $t_{1/2}$ values calculated on mineralization ($^{14}CO_2$ evolution) and parent-compound disappearance from laboratory studies. The $t_{1/2}$ values were smaller for field than for laboratory studies. Rao and Davidson (1980) attribute this to the multitude of factors that can affect pesticide disappearance in the field while only one factor is studied in the laboratory. Rao and Davidson (1982) suggested that pesticides be classified into three groups based on $t_{1/2}$ values (Table 6.5): nonpersistent ($t_{1/2} < 20$ days), moderately persistent ($20 \leq t_{1/2} \leq 100$ days), and persistent ($t_{1/2} > 100$ days). Most chlorinated hydrocarbons are grouped as persistent, while carboxylkanoic acid herbicides are nonpersistent. The *s*-triazines, substituted ureas, and carbamate pesticides are moderately persistent.

Pesticides in the first group are biodegradable and microbial decomposition is the major factor in degradation of the pesticides, but chemical degradation, photodecomposition, and volatilization may also be important (Rao and Davidson, 1980). The same processes are also important in degrading the pesticides in the second group above. In the latter group, metabolites may also accumulate due to incomplete degradation (Rao and Davidson, 1980). Mineralization rates of persistent pesticides are almost zero, and residue formation and metabolite accumulation in soils are significant.

TABLE 6.4 Degradation Rate Coefficients and Half-Lives ($t_{1/2}$) for Several Pesticides under Laboratory and Field Conditions[a]

		Rate coefficient (days^{-1})		$t_{1/2}$ (days)	
Pesticide	Condition	Mean	%CV	Mean	%CV
A. Herbicides					
2,4-D	Lab.[b]	0.066	74.2	16	56.3
	Lab.	0.051	23.5	15	33.3
	Field	3.6	83.3	5	100.0
2,4,5-T	Lab.	0.029	51.7	33	66.7
	Lab.[b]	0.035	82.9	16	68.8
Atrazine	Lab.[b]	0.019	47.4	48	68.8
	Lab.	0.0001	70.4	6900	71.5
	Field	0.042	33.3	20	50.0
Simazine	Lab.[b]	0.014	71.4	75	73.3
	Field	0.022	95.5	64	93.8
Trifluralin	Lab.[b]	0.008	65.5	132	82.6
	Lab.[b] (anaerobic)	0.025	—	28	—
	Lab. (chain)	0.0013	—	544	—
	Field	0.02	65.0	46	41.3
Bromacil	Lab.[b]	0.0077	49.4	106	42.5
	Lab.	0.0024	116.2	901	116.2
	Field	0.0038	100.0	349	76.8
Terbacil	Lab.[b]	0.015	33.3	50	26.0
	Lab.	0.0045	124.0	679	124.5
	Field	0.006	55.0	175	88.6
Linuron	Lab.[b]	0.0096	19.8	75	18.7
	Field	0.0034	41.2	230	29.3
Diuron	Lab.	—	—	—	—
	Field	0.0031	58.1	328	64.6
Dicamba	Lab.[b]	0.022	80.2	14	85.7
	Lab. (ring)	0.0022	—	309	—
	Lab. (chain)	0.0044	—	147	—
	Field	0.093	16.1	8	12.5
Picloram	Lab.[b]	0.0073	58.9	138	67.4
	Lab.	0.0008	111.3	8600	184.2
	Field	0.033	51.5	31	77.4
Dalapon	Lab.[b]	0.047	—	15	—
TCA	Lab.[b]	0.059	103.4	46	119.6
	Field	0.073	—	22	—
Glyphosate	Lab.[b]	0.1	121.0	38	139.5
	Lab.	0.0086	93.0	903	191.8
Paraquat	Lab.[b]	0.0016	—	487	—
	Field	0.00015	—	4747	—

(continued)

TABLE 6.4 (*Continued*)

Pesticide	Condition	Rate coefficient ($days^{-1}$)		$t_{1/2}$ (days)	
		Mean	%CV	Mean	%CV
B. Insecticides					
Parathion	Lab.[b]	0.029	48.3	35	82.9
	Field	0.057	101.8	18	44.4
Methyl parathion	Lab.[b]	0.16	—	4	—
	Field	0.046	—	15	—
Diazinon	Lab.[b]	0.023	108.7	48	62.5
	Field	0.022	—	32	—
Fonofos	Lab.[b]	0.012	—	60	—
Malathion	Lab.[b]	1.4	71.4	0.8	87.5
Phorate	Lab.[b]	0.0084	—	82	—
	Field	0.01	30.0	7.5	24.0
Carbofuran	Lab.[b]	0.047	87.2	37	94.6
	Lab.	0.0013	—	535	—
	Lab.[b] (anaerobic)	0.026	50.0	44	95.4
	Field	0.016	87.5	68	61.8
Carbaryl	Lab.[b]	0.037	56.8	22	40.9
	Lab (Chain)	0.0063	101.6	309	91.9
	Field	0.10	79.2	12	91.7
DDT	Lab.[b]	0.00013	130.8	1657	98.3
	Lab.[b] (anaerobic)	0.0035	82.9	692	123.4
Aldrin and Dieldrin	Lab.[b]	0.013	—	53	1
	Field	0.0023	100.0	1237	198.4
Endrin	Lab.[b] (anaerobic)	0.03	53.3	31	61.3
	Field (aerobic)	0.0015	—	460	—
	Field (anaerobic)	0.0053	—	130	—
Chlordane	Field	0.0024	104.2	1214	202.1
Heptachlor	Lab.[b]	0.011	—	63	—
	Field	0.0046	119.6	426	82.6
Lindane	Lab.[b]	0.0026	—	266	—
	Lab. (anaerobic)	0.0046	—	151	—
C. Fungicides					
PCP	Lab.[b]	0.02	60.0	48	60.4
	Lab. (anaerobic)	0.07	44.3	15	100.0
	Field	0.05	—	14	—
Captan	Field	0.231	—	3	—

[a] From Rao and Davidson (1982), with permission.

[b] These rates are based on the disappearance of solvent-extractable parent compound under aerobic incubation conditions, unless stated otherwise.

TABLE 6.5 Grouping of Pesticides Based on Their Persistence in Soils under Laboratory Incubation Conditions[a]

Nonpersistent, $t_{1/2} < 20$ days	Moderately persistent, $20 \leq t_{1/2} \leq 100$ days		Persistent, $t_{1/2} > 100$ days
2,4-D	Atrazine	Phorate	Trifluralin
2,4,5-T	Simazine	Carbofuran	Bromacil
Dicamba	Terbacil	Carbaryl	Picloram
Dalapon	Linuron	Aldrin	Paraquat
Methyl parathion	TCA	Dialdrin	DDT
Malathion	Glyphosate	Endrin	Chlordane
Captan	Parathion	Heptachlor	Lindane
	Diazinon	PCP	
	Fonofos		

[a] Persistence as determined by the rate of disappearance of the solvent-extractable parent compound under aerobic laboratory incubation conditions. From Rao and Davidson (1982), with permission.

REACTION RATES AND MECHANISMS OF ORGANIC POLLUTANTS

Besides pesticides, toxic organic substances are of great concern as we attempt to preserve the quality of our environment. Many of these substances have been deposited into aquatic and soil environments. In addition to understanding the equilibrium aspects of these pollutants in soils and sediments, it is imperative that there be an understanding of the rates and mechanisms of retention and mobility. Unfortunately, few of these studies have appeared in the scientific literature. This is most certainly an area of research in the soil and environmental sciences that needs extensive investigation.

Karickhoff (1980) and Karickhoff *et al.* (1979) have studied sorption and desorption kinetics of hydrophobic pollutants on sediments. Sorption kinetics of pyrene, phenanthrene, and naphthalene on sediments showed an initial rapid increase in sorption with time (5–15 min) followed by a slow approach to equilibrium (Fig. 6.7). This same type of behavior was observed for pesticide sorption on soils and soil constituents and suggests rapid sorption on readily available sites followed by tortuous diffusion-controlled reactions. Karickhoff *et al.* (1979) modeled sorption of the hydrophobic aromatic hydrocarbons on the sediments using a two-stage kinetic process. The chemicals were fractionated into a "labile" state (equilibrium occurring in 1 h) and a "nonlabile" state.

However, as has sometimes been seen with pesticides, the rates of

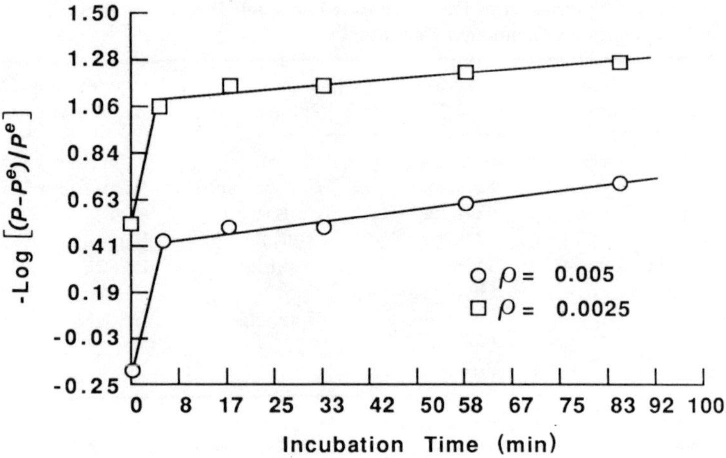

Figure 6.7. Phenanthrene sorption kinetics on a sediment, where ρ is the sediment/water ratio, P is the solution-phase pollutant concentration, and P^e is the equilibrium solution-phase concentration of the pollutant. [From Karickhoff (1980), with permission.]

sorption and desorption kinetics of organic pollutants on soils and sediments are quite different. For example, sorption of a hexachlorobiphenyl on sediments, clay minerals, and silica was characterized by rapid sorption (minutes to hours) but the desorption process was quite slow (DiToro and Harzempa, 1982). Karickhoff (1980) observed a significant difference in the extractability of sorbed organics depending on the time of equilibration or incubation. After short incubation periods (<5 min), >90% of the sorbed chemicals could be extracted from sediments with hexane for 3 min. However, after 3–5 h of incubation, the fraction of sorbed chemical extracted decreased to 0.5. These findings again point to a particle process where the organic chemical is slowly incorporated into either particle aggregates or sorbed components.

SUPPLEMENTARY READING

Bowman, B. T., and Sans, W. W. (1985). Partitioning behavior of insecticides in soil-water systems. II. Desorption hysteresis effects. *J. Environ. Qual.* **14,** 270–273.

Hance, R. J. (1967). Speed of attainment of sorption equilibria in some systems involving herbicides. *Weed Res.* **7,** 29–36.

Hance, R. J., ed. (1981). "Interactions Between Herbicides and the Soil." Academic Press, New York.

Haque, R., Lindstrom, F. T., Freed, V. H., and Sexton, R. (1968). Kinetic study of the sorption of 2,4-D on some clays. *Environ. Sci. Technol.* **2,** 207–211.

Karickhoff, S. W. (1980). Sorption kinetics of hydrophobic pollutants in natural sediments. *In* "Contaminants and Sediments: Analysis, Chemistry, Biology" (R. H. Baker, ed.), Vol 2, pp. 193–205. Ann Arbor Sci. Publ., Ann Arbor, Michigan.

Kearny, P. C., and Helling, C. S. (1969). Reactions of pesticides in soils. *Res. Rev.* **25,** 25–44.

Leenheer, J. A., and Ahlrichs, J. L. (1971). A kinetic and equilibrium study of the adsorption of carbaryl and parathion upon soil organic matter surfaces. *Soil Sci. Soc. Am. Proc.* **35,** 700–704.

Lindstrom, F. T., Haque, R., and Coshow, W. R. (1970). Adsorption from solution. III. A new model for the kinetics of adsorption-desorption processes. *J. Phys. Chem.* **74,** 495–502.

McCall, P. J., and Agin, G. L. (1985). Desorption kinetics of picloram as affected by residence time in the soil. *Environ. Toxicol. Chem.* **4,** 37–44.

7

Rates of Chemical Weathering

INTRODUCTION

The application of chemical kinetics to weathering has occurred only recently. Wollast (1967) was one of the first researchers to do this when he studied silica release from K-feldspars with time. Since his work, many studies have appeared, particularly in the geochemistry literature, on weathering of feldspars (see, e.g., Velbel, 1985) and of pyroxenes and amphiboles (Schott *et al.*, 1981), and on dissolution of oxides and aluminosilicates (Stumm *et al.*, 1985) and of calcite (Amhrein *et al.*, 1985). The rates of chemical weathering in soils and sediments depend on several factors, including mineralogy, temperature, flow rate, surface area, ligand and CO_2 concentration in soil water, and H^+ concentration (Stumm *et al.*, 1985). Chemical kinetic weathering is a broad subject, and in this chapter the focus will be on dissolution rates and mechanisms of feldspars, ferromagnesian minerals, oxides, and hydroxides.

RATE-LIMITING STEPS IN MINERAL DISSOLUTION

There are basically three rate-limiting mechanisms for mineral dissolution assuming a fixed degree of undersaturation. They are (1) transport of solute away from the dissolved crystal or transport-controlled kinetics,

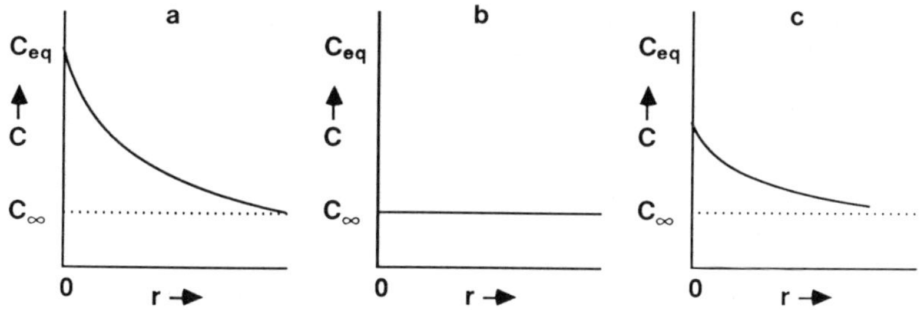

Figure 7.1. Rate-limiting steps in mineral dissolution: (a) transport-control, (b) surface reaction-control, and (c) mixed transport and surface reaction control. Concentration C versus distance r from a crystal surface for three rate-controlling processes and where C_{eq} is the saturation concentration and C_s is the concentration out in solution. [From Berner (1980), with permission.]

(2) surface reaction-controlled kinetics where ions or molecules are detached from the surface of crystals, and (3) a combination of transport and surface reaction-controlled kinetics (Berner, 1978). These three rate-limiting processes are schematically shown in Fig. 7.1.

In transport-controlled kinetics, the dissolution ions are detached very rapidly and accumulate to form a saturated solution adjacent to the surface. Then the dissolution is controlled by ion transport by advection and diffusion into the undersaturated solution. The rate of transport-controlled kinetics is affected by stirring and flow velocity. As they increase, transport and dissolution both increase (Berner, 1978).

With surface reaction-controlled kinetics, ion detachment is slow and ion accumulation at the crystal surface cannot keep up with advection and diffusion. In this type of phenomenon, the concentration level next to the crystal surface is tantamount to the surrounding solution concentration. Increased flow rate and stirring have no effect on the rate of surface reaction-controlled rate processes (Berner, 1978, 1983).

The third type of rate-limiting mechanism for mineral dissolution—mixed or partial surface reaction-controlled kinetics—exists when the surface detachment is fast enough that the surface concentration builds up to levels greater than the surrounding solution concentration but lower than that expected for saturation (Berner, 1978).

Dissolution occurring by a surface reaction is often slower than by transport-controlled kinetics because the latter results from more rapid surface detachment. There appears to be a good correlation between the solubility of a mineral and the rate-controlling mechanism for dissolution. Table 7.1 lists dissolution rate-controlling mechanisms for a number of substances. The less soluble minerals all dissolve by surface reaction-controlled kinetics. Silver chloride is an exception, but its dissolution

TABLE 7.1 Dissolution Rate-Controlling Mechanism for Various Substances Arranged in Order of Solubilities in Pure Water (Mass of Mineral That Will Dissolve to Equilibrium)[a]

Substance	Solubility (mol l^{-1})	Dissolution rate control
$Ca_5(PO_4)_3OH$	2×10^{-8}	Surface reaction
$KAlSi_3O_8$	3×10^{-7}	Surface reaction
$NAlSi_3O_8$	6×10^{-7}	Surface reaction
$BaSO_4$	1×10^{-5}	Surface reaction
AgCl	1×10^{-5}	Transport
$SrCO_3$	3×10^{-5}	Surface reaction
$CaCO_3$	6×10^{-5}	Surface reaction
Ag_2CrO_4	1×10^{-4}	Surface reaction
$PbSO_4$	1×10^{-4}	Mixed
$Ba(IO_3)_2$	8×10^{-4}	Transport
$SrSO_4$	9×10^{-4}	Surface reaction
Opaline SiO_2	2×10^{-3}	Surface reaction
$CaSO_4 \cdot 2H_2O$	5×10^{-3}	Transport
$Na_2SO_4 \cdot 10H_2O$	2×10^{-1}	Transport
$MgSO_4 \cdot 7H_2O$	3×10^{0}	Transport
$Na_2CO_3 \cdot 10H_2O$	3×10^{0}	Transport
KCl	4×10^{0}	Transport
NaCl	5×10^{0}	Transport
$MgCl_2 \cdot 6H_2O$	5×10^{0}	Transport

[a]From Berner (1980), with permission.

mechanism may be affected by photochemical alterations during the kinetic studies. Most of the minerals involved in weathering have solubilities in the lower range shown in Table 7.1, and it would appear that their dissolution is a surface reaction-controlled process.

FELDSPAR, AMPHIBOLE, AND PYROXENE DISSOLUTION KINETICS

Much attention has been given to feldspar dissolution kinetics over the past 20 years or so. This is largely attributable to feldspars being the most abundant minerals in igneous and metamorphic rocks. Their ubiquity in soils is also well known where they affect the resultant clay mineralogy and potassium status of soils. Another reason feldspar dissolution rates have been studied profusely has been the controversy over dissolution mechanisms. These arguments are discussed below, along with discussions on amphibole and pyroxene dissolution rates which are similar to feldspar dissolution.

Parabolic Kinetics

A number of workers (Wollast, 1967; Huang and Kiang, 1972; Luce *et al.*, 1972) have observed that feldspar weathering conforms to the parabolic diffusion law (Chapter 2). An example of this is shown in Fig. 7.2. Much research effort has gone into explaining why parabolic kinetics could be operational for feldspar dissolution. Several explanations have been given, and these are discussed below.

Protective Surface or Leached Layer. Correns and von Englehardt (1938) hypothesized that a protective surface layer was created as feldspars weather and that diffusion of dissolved products occurred through this surface layer, which became thicker with time. Garrels (1959) and Garrels and Howard (1959) believed that the surface layer was made up of H-feldspar, which resulted when H replaced alkali and alkaline earth cations from the feldspar surface. Wollast (1967) said it was composed of a aluminous or aluminosilicate precipitate. For a number of years a diffusion inhibiting surface layer was widely accepted as the main rate-controlling factor in the dissolution of not only feldspars, but also magnesium silicates (Luce *et al.*, 1972) and all noncarbonate and non-sulfur-bearing rock-forming minerals.

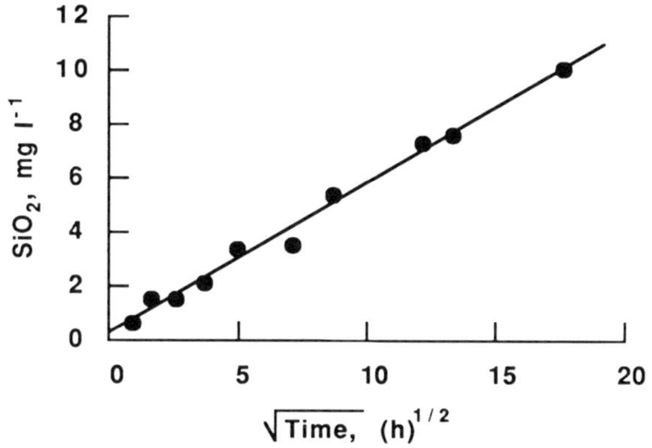

Figure 7.2. Example of parabolic kinetics showing linear behavior of silica concentration versus square root of time. Data from Wollast (1967). [From Velbel (1986), with permission.]

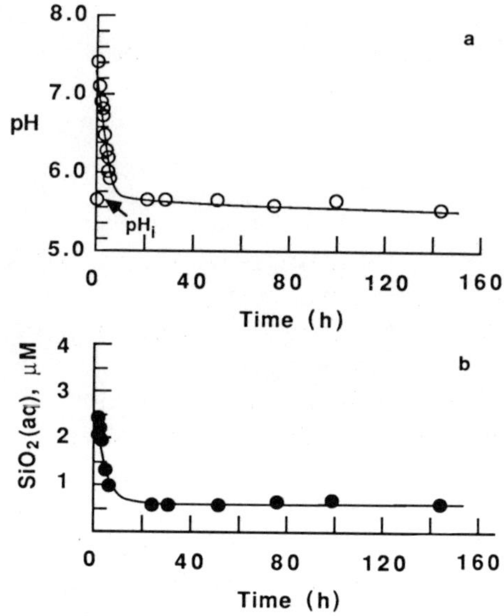

Figure 7.3. Changes in (a) pH, where pH_i is the pH of the input solution, and (b) the concentrations of Si with time for albite dissolution (100–200 μm size fractions) and pH 5.68 water as the input solution. [From Chou and Wollast (1984), with permission.]

The existence or nonexistence of a residual layer has been studied using surface chemistry techniques such as scanning electron microscopy (SEM) and X-ray photoelectron spectroscopy (XPS) and solution chemistry calculations. Nickel (1973) calculated the thickness of a residual layer on albite from the mass of dissolved alkalis and alkaline earths released during laboratory weathering. The surface area was also measured, and the thickness of the residual layer was found to range from 0.8 to 8 nm. These results suggested a very thin layer, which would not cause parabolic kinetics.

Chou and Wollast (1984, 1985) employed a fluidized-bed reactor to study albite dissolution with time. Figure 7.3 shows a short-term experiment run at room temperature and pressure using water as the input solution. There is a fast nonstoichiometric dissolution early in the reaction period that decreases rapidly until a steady state is approached. Linear kinetics and stoichiometric dissolution prevail later. If the pH of the input solution is changed, however, there is an increase in dissolution rate (Fig. 7.4) similar to the beginning of an experiment (Fig. 7.3).

Chou and Wollast (1984, 1985) concluded that the behavior in dissolution rate when pH was changed was due to the formation of a new surface

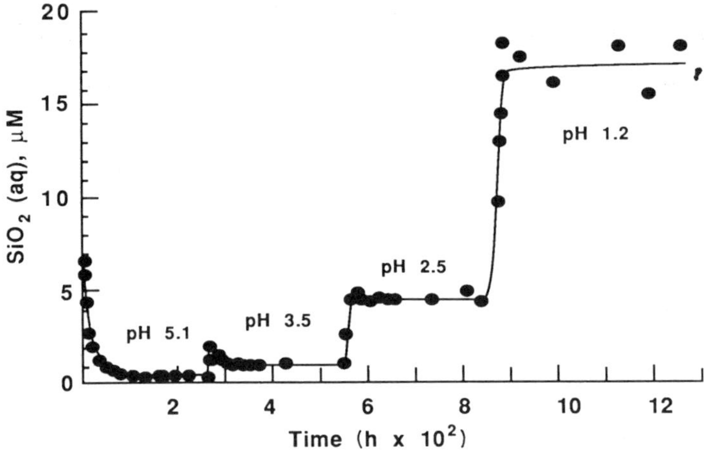

Figure 7.4 Changes in the concentration of Si under acidic conditions of pH 1.2–5.1 (input solution) for albite dissolution (50–100μm size fraction. [From Chou and Wollast (1984), with permission.]

layer, and diffusion-controlled processes were operable. The residual layer thickness based on a diffusion model was calculated to range from 25×10^{-8} cm at pH 6.9 to 70×10^{-8} cm at pH 1.2. Chou and Wollast (1984) also compared the thicknesses of the residual layer and the diffusion coefficients from a number of other studies (Table 7.2). There is a broad range in the thicknesses of the leached layer, which are dependent on experimental conditions.

It is important to note that the layer thicknesses reported above were based strictly on solution chemistry analyses. Several reports have appeared on the thicknesses of leached layers using surface chemistry techniques. Petrovic *et al.* (1976) used XPS and analyzed K, Al, and Si content of altered K-feldspar grains and found the leached layer was <1.7 nm. Layer thicknesses for dissolution of enstatite, diopside, and tremolite based on XPS data are reported in Table 7.3.

In the work of Schott *et al.* (1981), two kinds of layers were considered (Table 7.3). The first type was assumed to be completely depleted of either Ca or Mg. With the second type, a linear increase in cationic concentrations with depth was assumed. In either case, the layer thicknesses were only of atomic dimensions. Schott *et al.* (1981) also compared these layer thicknesses to those calculated based on solution chemistry analyses and mass balance considerations. Thicknesses of totally cation-depleted leached layers (pH=6) were 0.2 nm for Mg in enstatite, 1.7 nm for Ca in diopside, and 1.4 nm for Ca in tremolite.

TABLE 7.2 Comparison of the Thicknesses of the Leached Layer and the Diffusion Coefficients Calculated from the Literature at Room Temperature, Based on Diffusion Models[a]

Reference	Mineral	pH	Thickness of the leached layer (10^{-8} cm)	Diffusion coefficient, log *D* ($cm^2\ s^{-1}$)
Correns (1963)	K in K-feldspar	3	300	−18.9
Wollast (1967)	K in orthoclase	4	150	−18.8[b]
Luce *et al.* (1972)	Mg in Mg-silicates	3–10	—	−18 to −15
Paces (1973)	Na in albite	—	2	−21.5[c]
		—	11	−20.2[c]
		—	12	−19.8[c]
Busenberg and Clemency (1976)	Na in albite	5	—	−21.1
	K in orthoclase	5	—	−20.6
	K in microcline	5	—	−21.0
Chou and Wollast (1984)	Na in albite	6.9	25[d]	−20.0
		3.5	56[d]	−19.7
		1.2	70[d]	−18.4

[a] From Chou and Wollast (1984), with permission.
[b] Recalculated from the data of Wollast (1967).
[c] Extrapolated by Paces from the high temperature and pressure data of Lagache (1965) on the dissolution of albite.
[d] Calculated assuming a linear gradient across the leached layer.

TABLE 7.3 Calculated Thicknesses of Leached Layers Using XPS data[a,b]

	pH	Temperature (K)	Thickness of the layer (nm): Mg total depletion ($\lambda = 3$ nm)	Mg linear increase ($\lambda = 3$ nm)	Ca total depletion ($\lambda = 2$ nm)	Ca linear increase ($\lambda = 2$ nm)
Enstatite (28-day dissolution)	6	293	0.30	0.50	—	—
	6	313	0.30	0.60	—	—
	1	293	1.20	2.40	—	—
Diopside (22-day dissolution)	6	293	0	0	0.30	0.60
	6	313	0	0	0.50	1.00
	1	293	0.60	1.10	0.80	1.60
Tremolite (24-day dissolution)	6	293	0.15	0.30	0.30	0.60
	6	313	0.20	0.50	0.35	0.70
	1	293	0.40	0.80	0.40	0.80

[a] From Schott *et al.* (1981), with permission.
[b] The symbol λ represents the estimated mean free path for photoelectrons of the appropriate energy.

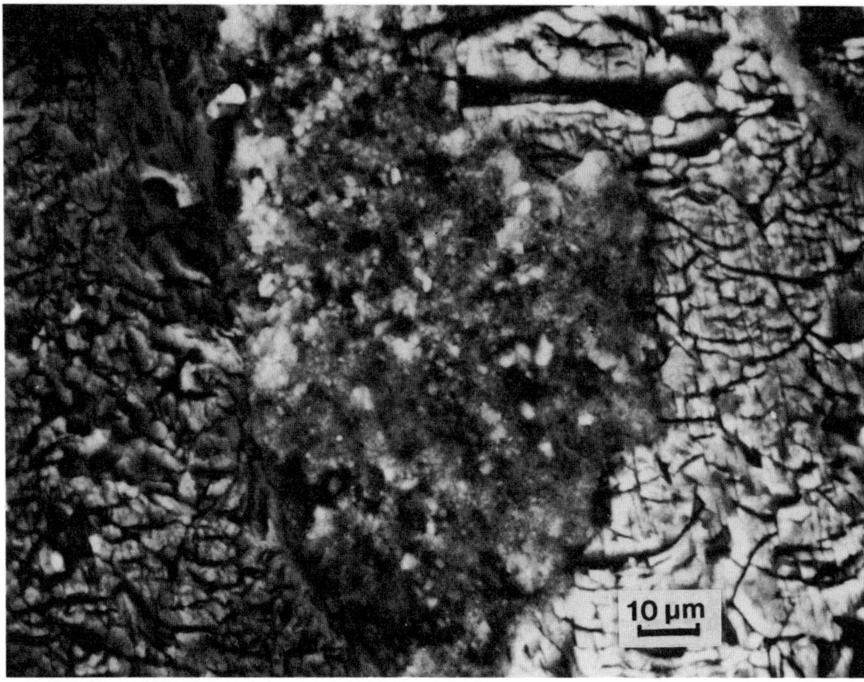

Figure 7.5. Scanning electron micrograph of a Kenansville soil feldspar showing etch pits. [From Sadusky *et al.* (1987), with permission.]

The controversy about the effect of a leached layer on parabolic kinetics still rages. Berner and co-workers (Berner and Holdren, 1977, 1979; Berner, 1978) have contended that feldspar and ferromagnesian dissolution is a surface reaction-controlled process. In addition to the XPS data cited above, which indicate a thin leached layer, they have used electron microscopy to show that etch pits (crystallographically controlled voids or negative crystals) exist in many laboratory and naturally weathered feldspars. An example of etch pits from a Delaware soil feldspar is shown in Fig. 7.5. Etch pits indicate the presence of dislocations in a crystal. These dislocations are extremely sensitive to selective attack or surface-controlled weathering. According to Berner (1978), the presence of etch pits indicates a surface-controlled reaction mechanism since minerals weathered through diffusion should show smooth, rounded surfaces. The smooth, rounded surfaces arise in part because ion detachment over the entire crystal surface is so rapid that selective etching does not occur.

A resolution to the discrepancies between leached layer thicknesses based on surface chemistry and dissolution studies has not been resolved.

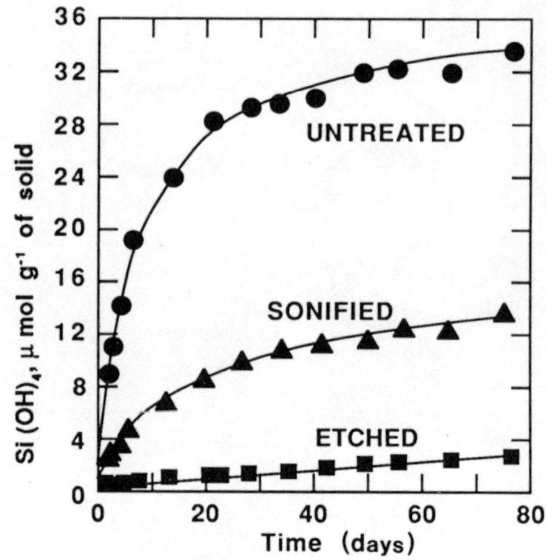

Figure 7.6. Silica release during the dissolution of untreated, sonified, and etched enstatite in buffer solution at pH = 6, T = 293 K. [From Schott *et al.* (1981), with permission.

However, Berner *et al.* (1985) suggest that the contradictions "may be illusory." They point out that limited incongruency could be accomodated with the XPS data sets Chou and Wollast (1984) dispute. Berner *et al.* (1985) suggest that spatially inhomogeneous incongruent dissolution along tubes (e.g., etch pits that penetrate deep into the interiors of mineral grains) could reconcile XPS measurements (to which the interior surfaces of tubes would not be accessible) with the incongruency and material balance requirements of laboratory studies, while still allowing surface-reaction control. Chou and Wollast (1985) tend to agree with Berner *et al.* (1985) concerning the role of dissolution-void interiors in resolving the above differences.

Structural Mineral Distortions and Hyperfine Particles. The early, high-rate stage observed for dissolution of feldspars and ferromagnesian minerals has also been ascribed to a structurally distorted or strained outer layer of the mineral, or to large quantities of hyperfine particles adhering electrostatically to the outer surface of the mineral (Velbel, 1986). An example of how pretreatment of minerals affects silica release from enstatite is shown in Fig. 7.6. The rate of silica dissolution of untreated and sonified samples was fast initially and then decreased with time, resulting in parabolic kinetics. This behavior was ascribed to initial rapid dissolution of ultrafine, supersoluble particles adhering to large grains that were produced during grinding (Holdren and Berner, 1979; Schott *et al.*, 1981).

However, when the enstatite sample was first pretreated with HF–H_2SO_4 to remove ultrafine particles and surface defects, there was a constant release of silica with time (linear kinetics).

Nonlinear Precipitation of Secondary Minerals from Solution. Most of the studies on dissolution of feldspars, pyroxenes, and amphiboles have employed batch techniques. In these systems the concentration of reaction products increases during an experiment. This can cause formation of secondary aluminosilicate precipitates and affect the stoichiometry of the reaction. A buildup of reaction products alters the ion activity product (IAP) of the solution vis-à-vis the parent material (Holdren and Speyer, 1986). It is not clear how secondary precipitates affect dissolution rates; however, they should depress the rate (Aagaard and Helgeson, 1982) and could cause parabolic kinetics. Holdren and Speyer (1986) used a stirred-flow technique to prevent buildup of reaction products.

Solution Composition. Changing solution composition can also cause apparent or true parabolic dissolution kinetics through the influence of changing pH and CO_2 equilibria or through an effect on chemical affinity and reverse rate (Helgeson *et al.*, 1984).

Dissolution Mechanism

In summary, the mechanism for dissolution of feldspars, pyroxenes, and amphiboles appears to involve a rapid hydrogen exchange for alkalis and alkaline earths, which creates a thin layer of hydrolyzed aluminosilicate. This residual layer ranges in thickness from several to a few tens of nanometers and is responsible for the initial nonstoichiometric release of alkalis and alkaline earths relative to Si and Al (Velbel, 1985).

Following this step there is continued dissolution, which removes whatever hyperfine particles may have resulted during sample preparation. After removing these, further dissolution breaks down the outer surface of the residual layer at the same rate that alkalis are replaced by hydrogen at the interface between fresh mineral surfaces and the residual layer. This releases all constituents to the solution. Release is now stoichiometric, based on solution chemistry and surface morphological results. Thus, the reaction is surface-controlled (Velbel, 1985).

However, as Velbel (1985) stated so well, more research is needed to see whether the above mechanism is operational under field conditions. We know that rates of mineral weathering are often lower in the field than in the laboratory. Natural feldspar weathering may be one to three times slower in the field (Paces, 1973). Velbel (1985) found that the rate of plagioclase weathering was about one order of magnitude slower in nature than in the laboratory. These results can be partly attributed to the differences in mineral surfaces weathered in the laboratory and in the field. For example, experimentally weathered feldspars are often fresh, rough, and characterized by kinks, ledges, and terraces, which result in higher dissolution rates than naturally weathered feldspars. The latter are often smooth and rounded and their surfaces may be partially covered with weathering products, which may inhibit dissolution.

DISSOLUTION RATES OF OXIDES AND HYDROXIDES

Several studies have been conducted on the rates of dissolution of oxides. The work of Stumm and coworkers is noteworthy in this area (Stumm *et al.*, 1983, 1985; Zutic and Stumm, 1984; Stumm, 1986). They have studied the effects of H^+ and various complex-forming anions on oxide dissolution rates and found that dissolution rate (v) depends strongly on the relative concentrations of proton surface groups $\{\exists\text{-}OH_2^+\}$ and ligand surface complexes $\{\exists\text{-}L\}$ such that (Stumm *et al.*, 1985)

$$\begin{aligned} v_H &\propto [\exists\text{-Me-}OH_2^+]^z \\ v_L &\propto [\exists\text{-Me-L}] \end{aligned} \tag{7.1}$$

where z is the valence of the metal in the metal oxide, Me is some metal, v_H is the surface proton-dependent rate, and v_L is the ligand surface complex rate.

Thus, the dissolution rate of oxides and hydroxides is composed of two additive rates such that

$$v = v_H + v_L = k_H[\exists\text{-}OH_2^+]^z + k_L[\exists\text{-L}] \tag{7.2}$$

Figure 7.7 shows how acids and ligands affect oxide surface groups and polarize the Me–O bonding in dissolution phenomena. In a ligand exchange reaction, the nucleophilic ligand L binds a Me center and replaces on OH group. This additionally polarizes the particular Me–O bonds (Stumm *et al.*, 1983).

Figure 7.8 shows schematically the various steps in the surface reaction-controlled dissolution of hydrous oxides or hydroxides. A number of protons equal to the valency of the Me center is needed to detach an Me(aq) group into the aqueous solution. Stepwise protonation occurs relatively quickly. The slow, rate-determining step is detachment of the Me(aq) group (Stumm, 1986). With a divalent metal oxide, two neighboring surface groups must become protonated. The rate of detachment (step

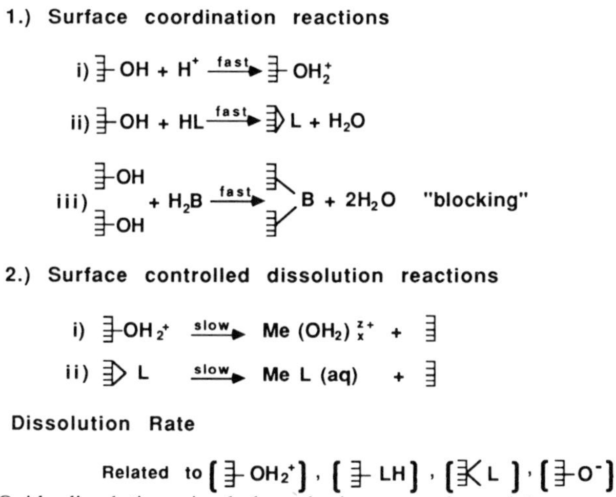

Figure 7.7. Oxide dissolution: simple hypothesis on rate-determining step. [From Stumm (1986), with permission.]

c) depends on the surface concentration of surface metal centers that contain two neighboring protonated surface hydroxo groups.

Stumm (1986) defines the probability of finding two neighboring protonated groups as being proportional to θ_H^2, where θ_H is the degree of surface protonation such that

$$\theta_H = \frac{[\exists\!-\!OH_2^+]}{[\exists\!-\!OH] + [\exists\!-\!OH_2^+]} \tag{7.3}$$

For a trivalent metal oxide, three neighboring surface sites must be protonated and the probability of finding them is $\propto \theta_H^3$.

The rate-determining step for oxide dissolution in a slightly acidic solution (reaction c in Fig. 7.8) is given in Eq. (7.4) and occurs at kink or step sites on the oxide or hydroxide surface. It can be expressed as

$$\nu_H = k_H \theta_H^z = k_H[\exists\!-\!OH_2^+]^z \tag{7.4}$$

For Al_2O_3, Eq. (7.4) could be expressed as

$$\nu_H = k\theta_H^3 = k_H[\exists\!-\!OH_2^+]^3 \tag{7.5}$$

Correspondingly, the ligand-catalyzed reaction rate can be expressed in terms of a term expressing the relative coverage of site with L, θ_L, such that

$$\nu_L = k_L \theta_L = k_L[\exists\!-\!L] \tag{7.6}$$

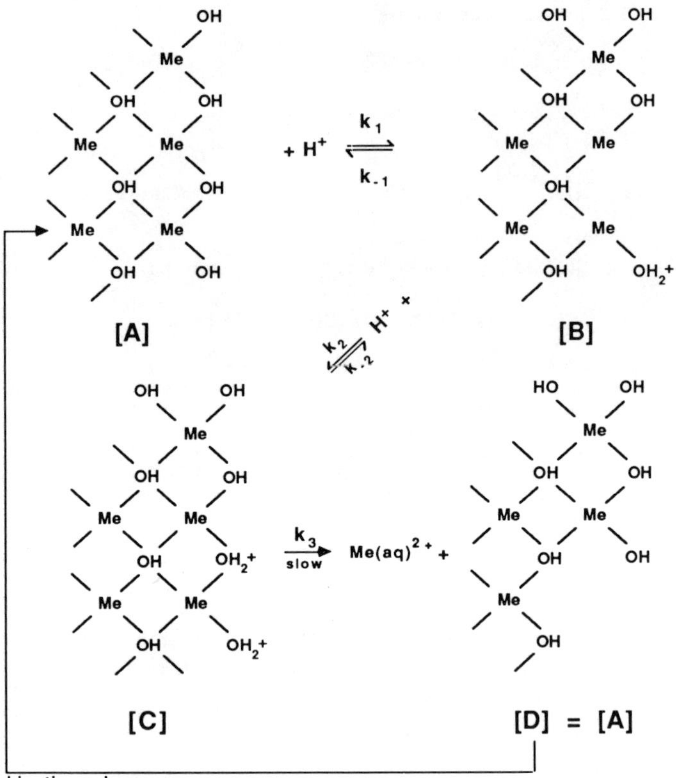

kinetic scheme:

$$A + H^{+} \underset{k_{-1}}{\overset{k_1}{\rightleftharpoons}} B \qquad \text{(fast)} \qquad \text{(a)}$$

$$B + H^{+} \underset{k_{-2}}{\overset{k_2}{\rightleftharpoons}} C \qquad \text{(fast)} \qquad \text{(b)}$$

$$C + H_2O \xrightarrow{k_3} A + Me\cdot aq^{2+} \qquad \text{(slow)} \qquad \text{(c)}$$

$$\frac{d[Me(aq)^{2+}]}{dt} \cong k_3\ [C] \quad ; \quad k_3 < k_1, k_2, k_{-2}$$

$$[C] = \text{prop}\ \theta_H^2 \qquad \theta_H = \text{degree of surface protonation}$$

Figure 7.8. Schematic representation of the steps involved in the dissolution of a (hydr)-oxide, illustrating that the rate-determining step is the detachment of a suitably protonated surface group. [From Stumm (1986), with permission.]

and the rate-determining step is detachment of the MeL complex (Fig. 7.9).

The effect of $[H^+]$ and $[\exists\!-\!OH_2]$ on dissolution of $\delta - Al_2O_3$ is shown in Fig. 7.10. One can see that the rate of dissolution is directly dependent on these two factors. The effect of ligands on dissolution rate is shown in Fig. 7.11. Here a linear relationship is seen between v_L and $[\exists\!-\!L]$. Stumm

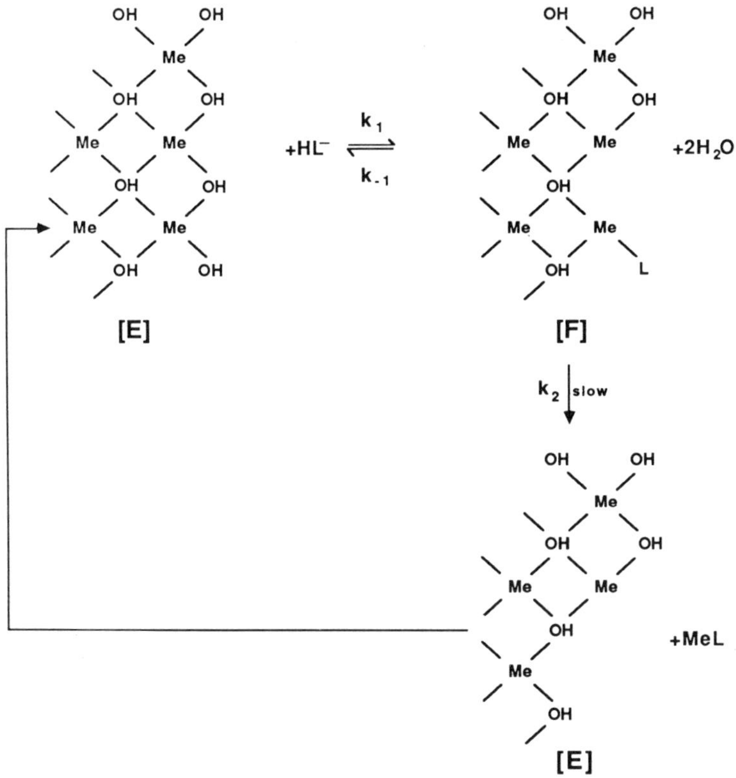

$$E + HL \underset{k_{-1}}{\overset{k_1}{\rightleftharpoons}} F + H_2O \qquad \text{(fast)} \qquad \text{(d)}$$

$$F + H_2O \xrightarrow{k_2} MeL + E \qquad \text{(slow)} \qquad \text{(e)}$$

$$\frac{d[MeL]}{dt} \cong k_2\,[F] \qquad [F] = \text{prop}\ \theta_L$$

Figure 7.9. In a ligand-catalyzed reaction, the surface complexation with a bidentate ligand is followed by the rate-determining detachment of the metal ligand complex. [From Stumm (1986), with permission.]

et al. (1985) have shown that the highest dissolution rate occurs when bidentate mononuclear surface chelates are foremost, such as oxalate (five-ring) and malonate and salicylate (six-ring compounds). The k_L values for these three chelates have been reported as 10.8, 6.9, and $12.5 \times 10^{-3}\ h^{-1}$, respectively. Oxalate also causes enhanced dissolution of $Fe(OH)_3$ (amorphous), $\alpha - Fe_2O_3$ (hematite), and α-FeOOH (goethite) (Stumm *et al.*, 1985).

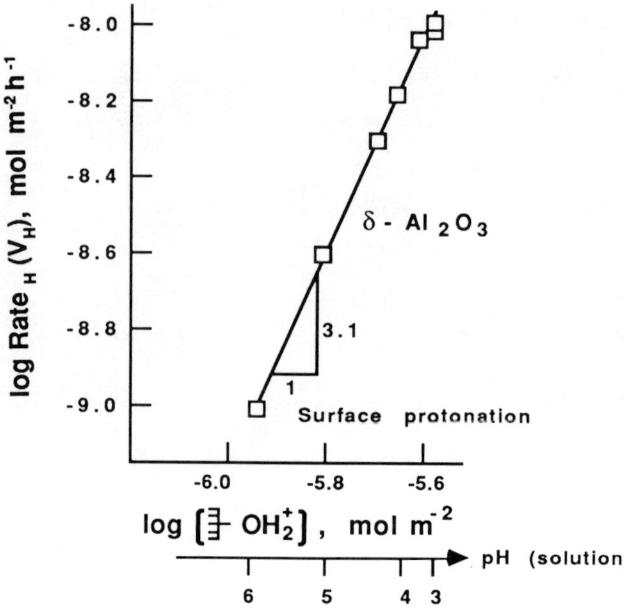

Figure 7.10 Effect of pH on the dissolution rate of $\delta - Al_2O_3$. This dependence can be reinterpreted in terms of a dependence on the concentration of protonated surface groups, [⫤OH_2^+]. The rate depends on [⫤OH_2^+]^3. [From Stumm *et al.* (1985), with permission.]

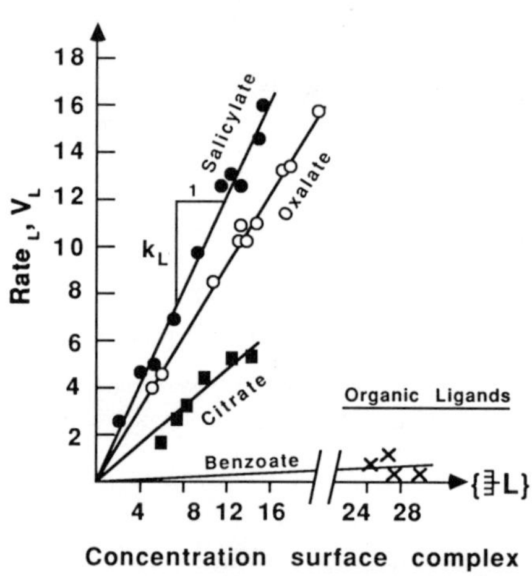

Figure 7.11. Dissolution rate dependence on the presence of organic ligand anions (pH 2.5–6) can be interpreted as a linear dependence on the surface concentrations of deprotonated ligands, [⫤L]; ν_L (nmol m^{-2} h^{-1}) is that portion of the rate that is dependent on surface complexes only. In the case of citrate and salicylate, at pH 4.5 corrections accounting for the protonation of the surface complexes were made. [From Stumm *et al.* (1985), with permission.]

TABLE 7.4 Hydrolysis Reaction Order for the Dissolution of Minerals[a]

Mineral	Formula	Solution	Reaction order, n
Dolomite	$(Ca,Mg)CO_3$	HCl	$[H^+]^{0.5}$
Bronzite	$(Mg,Ca)SiO_3$	HCl	$[H^+]^{0.5}$
Enstatite	$MgSiO_3$	HCl	$[H^+]^{0.8}$
Diopside	$CaMgSi_2O_6$	HCl	$[H^+]^{0.7}$
K-feldspar	$KAlSi_3O_8$	Buffer	$[H^+]^{0.33}$
Iron hydroxide	$Fe(OH)_3$ gel	Various acids	$[H^+]^{0.48}$
Aluminum oxide	γ-Al_2O_3	HCl	$[H^+]^{0.4}$
Gibbsite	$Al(OH)_3$	HNO_3, H_2SO_4	$[H^+]^{1.0}$[b]

[a] From Stumm *et al.* (1985), with permission.
[b] Bloom and Erich (1987).

Dissolution of oxides and hydroxides as well as several other minerals in acids is usually of fractional order (Table 7.4).

However, Bloom and Erich (1987) found that in NO_3^- and SO_4^{2-} solutions, the dissolution reaction for gibbsite was first-order with respect to $[H^+]$ below pH 2.5. In phosphate solutions, there was no dependence of the rate of dissolution on $[H^+]$. They also found that dissolution rate (v) was affected by the concentration of NO_3^-, SO_4^{2-}, and PO_4^{3-}. The order dependency of v on NO_3^-, SO_4^{2-}, and PO_4^{3-} concentration was 0.56, 0.36, and 0.88, respectively.

As mentioned earlier, the dissolution of oxides and hydroxides, like feldspars and ferromagnesian minerals, appears to be a surface-controlled reaction. One indication of this is the high E values found by several investigators. Bloom and Erich (1987) obtained E values ranging from 59 ± 4.3 to 67 ± 0.6 kJ mol^{-1} for gibbsite dissolution in acid solutions (pH 1.5–4.0). These values are much higher than for diffusion-controlled reactions reported earlier.

SUPPLEMENTARY READING

Berner, R. A. (1978). Rate control of mineral dissolution under earth surface conditions. *Am. J. Sci.* **278,** 1235–1252.

Berner, R. A. (1983). Kinetics of weathering and diagenesis. *Rev. Mineral.* **8,** 111–134.

Helgeson, H. C., Murphy, W. M., and Aagaard, P. (1984). Thermodynamic and kinetic constraints on reaction rates among minerals and aqueous solutions. II. Rate constants, effective surface area, and the hydrolysis of feldspar. *Geochim. Cosmochim. Acta* **48,** 2405–2432.

Holdren, G. R., Jr., and Speyer, P. M. (1986). Stoichiometry of alkali feldspar dissolution at room temperature and various pH values. *In* "Rates of Chemical Weathering of Rocks and Minerals" (S. M. Colman and D. P. Dethier, eds.), pp. 61–81. Academic Press, Orlando, Florida.

Schott, J., and Berner, R. A. (1985). Dissolution mechanisms of pyroxenes and olivines during weathering. *In* "The Chemistry of Weathering" (J. I. Drever, ed.), pp. 35–53. Reidel Publ., Dordrecht, The Netherlands.

Schott, J., Berner, R. A., and Lennart-Sjöberg, E. L. (1981). Mechanism of pyroxene and amphibole weathering. I. Experimental studies of Fe-free minerals. *Geochim. Cosmochim. Acta* **45,** 2123–2135.

Stumm, W. (1986). Coordinative interaction between soil solids and water. An aquatic chemist's point of view. *Geoderma* **38,** 19–30.

Stumm, W., Furrer, G., and Kunz, B. (1983). The role of surface coordination in precipitation and dissolution of mineral phases. *Croat. Chem. Acta* **56,** 593–611.

Stumm, W., Furrer, G., Wieland, E., and Zinder, B. (1985). The effects of complex-forming ligands on the dissolution of oxides and aluminosilicates. *In* "The Chemistry of Weathering" (J. I. Drever, ed.), pp. 55–74. Reidel Publ., Dordrecht, The Netherlands.

Velbel, M. A. (1986). Influence of surface area, surface characteristics, and solution composition on feldspar weathering rates. *ACS Symp. Ser.* **323,** 615–634.

Wollast, R. (1967). Kinetics of the alteration of K-feldspar in buffered solutions at low temperature. *Geochim. Cosmochim. Acta* **31,** 635–648.

Wollast, R., and Chou, L. (1985). Kinetic study of the dissolution of albite with a continuous flow through fluidized bed reactor. *In* "Chemistry of Weathering" (J. I. Drever, ed.), pp. 75–96. Reidel Publ., Dordrecht, The Netherlands.

Zutic, V., and Stumm, W. (1984). Effect of organic acids and fluoride on the dissolution kinetics of hydrous alumina. A model study using the rotating disc electrode. *Geochim. Cosmochim. Acta* **48,** 1493–1503.

8

Redox Kinetics

INTRODUCTION

The role of transition metal oxide/hydroxide minerals such as Fe and Mn oxides in redox reactions in soils and aqueous sediments is pronounced (Stumm and Morgan, 1980; Oscarson *et al.*, 1981a). These oxides occur widely as suspended particles in surface waters and as coatings on soils and sediments (Taylor and McKenzie, 1966).

It is well documented that reductive dissolution of oxide/hydroxide minerals of Mn(III/IV), Fe(III), Co(III), and Pb(IV), which are thermodynamically stable in oxygenated solutions at neutral pH, are reduced to divalent metal ions under anoxic conditions when reducing agents are present (Stone, 1986). Alterations in the oxidation states of these metals greatly change their solubility and thus mobility in soil and aqueous environments. For example, reduction of Fe(III) to Fe(II) increases Fe solubility vis-à-vis the oxide/hydroxide phase by eight orders of magnitude (Stumm and Morgan, 1981).

Reductive dissolution of transition metal oxide/hydroxide minerals can be enhanced by both organic and inorganic reductants (Stone, 1986). There are numerous examples of natural and xenobiotic organic compounds that are efficient reducers of oxides and hydroxides. Organic reductants associated with carboxyl, carbonyl, phenolic, and alcoholic functional groups of soil humic materials are one example. However, large differences exist in their redox activities. Other organic reductants include microorganisms in soils and sediments. For example, it has been shown that two microbial metabolites, oxalate and pyruvate, reduce and dissolve Mn(III/IV) oxide particles at considerable rates (Stone, 1987a). Degradation of toxic organics in soil and aqueous environments by Mn(III/IV) oxides could have significant effects on ameliorating environmental quality.

Another very important role that metal oxides such as Mn(III/IV) play in soils and sediments is the oxidation of inorganic cations. These reactions can be both advantageous and deleterious to environmental quality. On the positive side, oxidation of toxic arsenite [As(III)] to arsenate [As(V)] by Mn(III/IV) oxides has been demonstrated (Oscarson *et al.*, 1980). On the negative side, Mn(III/IV) oxides can effect oxidation of Cr(III) and Pu(III) to Cr(VI) and Pu(VI). These latter forms are very mobile in soils; consequently, they can be toxic pollutants in the underlying aquatic environment (Amacher and Baker, 1982).

In this chapter, reductive dissolution rates and mechanisms of oxides will be discussed. Additionally, oxidation kinetics of As(III), Cr(III), and Pu(III) by Mn(III/IV) oxides will covered.

REDUCTIVE DISSOLUTION OF OXIDES BY ORGANIC REDUCTANTS

Reaction Scheme and Mechanism

The reductive dissolution of metal oxides such as Mn(III/IV) oxides by organic reductants occurs by the following sequential steps (Stone, 1986): (1) diffusion of reductant molecules to the oxide surface, (2) surface chemical reaction, and (3) diffusion of reaction products from the oxide surface. Steps (1) and (3), which are transport steps, are influenced by both the interfacial concentration gradient and the electrical potential gradient due to the net charge of the oxide surface.

The rate-controlling step in reductive dissolution of oxides is surface chemical reaction control. The dissolution process involves a series of ligand-substitution and electron-transfer reactions. Two general mechanisms for electron transfer between metal ion complexes and organic compounds have been proposed (Stone, 1986): inner-sphere and outer-sphere. Both mechanisms involve the formation of a precursor complex, electron transfer with the complex, and subsequent breakdown of the successor complex (Stone, 1986). In the inner-sphere mechanism, the reductant

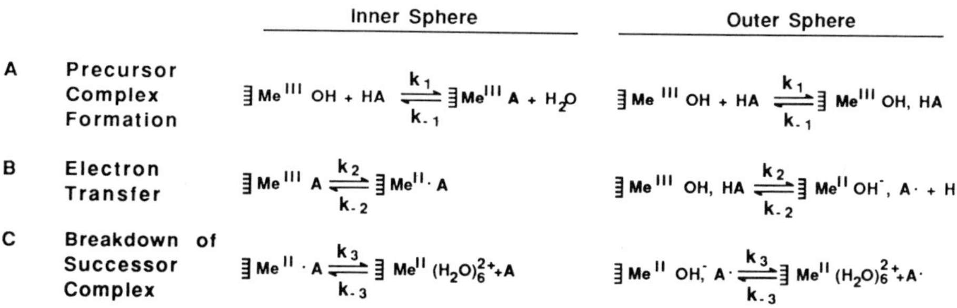

Figure 8.1. Reduction of tervalent metal oxide surface sites by phenol (HA) showing inner-sphere and outer-sphere mechanisms. [From Stone (1986), with permission.]

enters the inner coordination sphere by way of ligand substitution and bonds directly to the metal center before electron transfer (Stone, 1986). In the outer-sphere mechanism, the inner coordination sphere is left intact and electron transfer is aided by an outer-sphere precursor complex (Stone, 1986). Both of the mechanisms can occur in parallel, and the overall reaction is dominated by the fastest pathway. An illustration of inner-sphere and outer-sphere mechanisms for reduction of tervalent metal oxide surface sites by phenol is shown in Fig. 8.1.

The oxidant strength of transition metal oxides usually decreases in the order $Ni_3O_4 > MnO_2 > MnOOH > CoOOH > FeOOH$. The rate of reductive dissolution in sediments and natural waters follows a similar order.

The kinetics of the reductive dissolution mechanisms shown in Fig. 8.1 can be derived using the principle of mass action. The kinetic expression for precursor complex formation by way of an inner-sphere mechanism (Stone, 1986) is

$$\frac{d[\exists \mathrm{Me(III)A}]}{dt} = k_1[\exists \mathrm{Me(III)OH}][\mathrm{HA}] - (k_{-1} + k_2)[\ \mathrm{Me(III)A}] + k_{-2}[\exists \mathrm{Me(II)A}] \tag{8.1}$$

For steady-state precursor complex concentration and assuming negligible back reactions for B and C, the rate of Me^{2+}(aq) formation is (Stone, 1986):

$$\frac{d[\mathrm{Me}^{2+}]}{dt} = \frac{k_1 k_2}{(k_{-1} + k_2)}[\exists \mathrm{Me(III)OH}][\mathrm{HA}](1 - e^{-k_3 t}) \tag{8.2}$$

An analogous rate expression can be written for the outer-sphere mechanism. From Eq. (8.2), it can be predicted that high rates of reductive dissolution are enhanced by high rates of precursor complex formation (large k_1), low desorption rates (small k_{-1}), high electron transfer rates (large k_2), and high rates of product release (high k_3).

It should also be pointed out that the rate of each of the reaction steps (precursor complex formation, electron transfer, and breakdown of successor complex) is affected by the chemical characteristics of the metal oxide surface sites and the nature of the reductant molecules. These aspects are discussed in detail in an excellent review by Stone (1986), and the reader is encouraged to refer to this article.

Specific Studies

A number of studies have appeared in the literature on reductive dissolution of Mn(III/IV) oxides, particularly by organic reductants. Rates of reductive dissolution by hydroquinone (Stone and Morgan, 1984a), substituted phenols (Stone, 1987b), and other organic reductants (Stone and Morgan, 1984b) have been determined.

Stone (1987a) studied reductive dissolution of Mn oxides by oxalate and pyruvate; an example with 1.00×10^{-4} *M* oxalate is shown in Fig. 8.2. The rate of dissolution was directly proportional to organic reductant concentration and increased as pH decreased. For oxalate, the rate of reductive dissolution at pH 5.0 was 27 times faster than at pH 6.0. For pyruvate, a similar change in pH increased the rate only by a factor of 3.

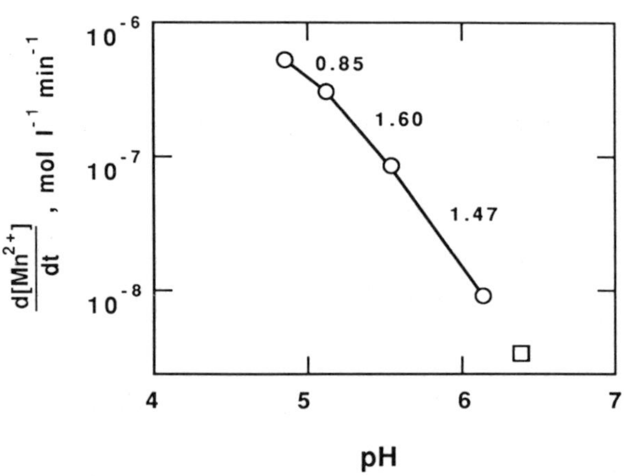

Figure 8.2. Rates of manganese oxide reductive dissolution by 1.00×10^{-4} *M* oxalate as a function of pH. Reactions were performed in 5.0×10^{-2} *M* NaCl using either acetate (○) or constant $-P_{CO_2}$ (□) buffers. ($[MnOx]_0$ is 4.81×10^{-5} *M*.) Numerical values are apparent reaction orders with respect to $[H^+]$. [From Stone (1987a), with permission.]

A similar effect of pH on dissolution rates of Mn(III/IV) oxides was observed by Stone (1987b) with substituted phenols. In this study, phenols with alkyl, alkoxy, or other electron, donating substituents were more slowly degraded. Stone (1987b) even found that *p*-nitrophenol, the most resistant phenol studied, reacted slowly with Mn(III/IV) oxides.

The implications of these results are extremely important, as they show that abiotic oxidation by Mn(III/IV) oxides can be a degradation mechanism for substituted phenols, which are so deleterious to environmental quality.

Sone (1987b) has attributed the pH dependence to one or a combination of two effects: protonation reactions that enhance the creation of precursor complexes, or increases in the protonation level of surface precursor complexes that increase rates of electron transfer.

The apparent order of reductive dissolution of Mn(III/IV) oxides can be determined experimentally from the slope of the log of reaction rate plotted versus pH (Stone, 1987a):

$$d[\mathrm{Mn}^{2+}]/dt = k[\mathrm{H}^+]^{\alpha}[\mathrm{reductant}]^{\beta} \tag{8.3}$$

$$\log(d[\mathrm{Mn}^{2+}]/dt) = \log k + \alpha \log[\mathrm{H}^+] + \beta \log[\mathrm{reductant}] \tag{8.4}$$

As can be seen from Fig. 8.2 for oxalate and, although not shown, also for pyruvate, the apparent order with respect to $[H^+]$, α, decreased with pH. Rates of reductive dissolution of Mn(III/IV) oxides by phenols were characterized by α values that also decreased with pH and were >1.0 at pH > 6.0 but were <0.5 as pH 4 was approached (Stone, 1987a).

Specifically adsorbed cations and anions may lower reductive dissolution rates by blocking oxide surface sites or by effecting release of Mn(II) into solution. Stone and Morgan (1984a) found that PO_4^{3-} considerably inhibited the reductive dissolution of Mn(III/IV) oxides by hydroquinone. For example, addition of 10^{-2} *M* PO_4^{3-} at pH 7.68 resulted in the dissolution rate being only 25% of the rate in the absence of PO_4^{3-}. The dissolution rate was affected more by PO_4^{3-} than by Ca^{2+}.

OXIDATION RATES OF CATIONS BY Mn(III/IV) OXIDES

Oxidation Kinetics of As(III)

Recently there has been much concern over arsenic in aquatic and soil environments from sources such as arsenical pesticides, smelters, coal-fired plants, and erosion caused by intensive land use (Huang and Liaw, 1979). Arsenic can exist in several oxidation states. It has been shown that in marine environments, As(V) can be reduced to As(III) by bacteria

(Johnson, 1972) and phytoplankton (Andreae and Klumpp, 1979). Johnson and Pilson (1975) showed that As(III) could be oxidized to As(V) in seawater.

Manganese(III/IV) oxides also play an important role in oxidizing cations such as As(III) to As(V) (Oscarson *et al.*, 1980, 1981a,b). This is a very beneficial effect, since As(III) is an extremely toxic pollutant and is more soluble and mobile (Deuel and Swoboda, 1972) than As(V). Iron(III) oxides in soils and sediments may also play a role in oxidizing As(III), but the kinetics of this reaction is very slow (Oscarson *et al.*, 1980). Phyllosilicates and calcite are not effective in oxidizing As(III) to As(V) (Oscarson *et al.*, 1981b).

When As(III) ($HAsO_2$) is added to MnO_2, it can either be oxidized to As(V) (H_3AsO_4) by MnO_2 (Oscarson *et al.*, 1981a) or sorbed on the oxide surface. The oxidization of As(III) by MnO_2 can be shown (Oscarson *et al.*, 1983a) as

$$HAsO_2 + MnO_2 = (MnO_2) \cdot HAsO_2 \tag{8.5}$$

$$(MnO_2) \cdot HAsO_2 + H_2O = H_3AsO_4 + MnO \tag{8.6}$$

$$H_3AsO_4 = H_2AsO_4^- + H^+ \tag{8.7}$$

$$H_2AsO_4^- = HAsO_4^{2-} + H^+ \tag{8.8}$$

$$(MnO_2) \cdot HAsO_2 + 2H^+ = H_3AsO_4 + Mn^{2+} \tag{8.9}$$

The first step is formation of an adsorbed layer [Eq. (8.5)]. With oxygen transfer, $HAsO_2$ is oxidized to H_3AsO_4 [Eq. (8.6)]. At pH ≤ 7, the major As(III) species is arsenious acid ($HAsO_2$), but the oxidation product H_3AsO_4 should dissociate to form equal quantities of $H_2AsO_4^-$ and $HAsO_4^{2-}$ with little H_3AsO_4 present at equilibrium [Eqs. (8.7) and(8.8)]. Thus, each mole of As(III) oxidized releases about 1.5 moles H^+ into the system. Assuming no other reaction takes place, pH will be lowered but will remain ~7.0. The H^+ produced after dissociation of H_3AsO_4 reacts with the adsorbed $HAsO_2$ on MnO_2, forming H_3AsO_4, and leads to the reduction and dissolution of Mn [Eq. (8.9)]. Therefore, every mole of As(II) that is oxidized to As(V) results in a mole of Mn(IV) in the solid phase being reduced to Mn(II) and partially dissolved in solution (Oscarson *et al.*, 1981a). Hydrogen ions dissociated from H_3AsO_4 can attack coatings of Al and Fe oxides and $CaCO_3$. If partial dissolution of the surface coatings occurs, fresh Lewis acid sites on MnO_2 are exposed and oxidation would be enhanced (Oscarson *et al.*, 1983a).

Oscarson *et al.* (1983a) found that the kinetics of As(III) depletion (oxidation plus sorption) by uncoated MnO_2 or Mn coated with Al or Fe oxides or $CaCO_3$ involved two rates: one before and one after 30 min. The kinetics conformed to first-order kinetics. Arsenite depletion kinetics decreased significantly as the degree of coatings increased (Table 8.1). This

TABLE 8.1 Rate Coefficients for the Depletion of As(III) by Untreated MnO_2 and MnO_2 Coated with Fe and Al Oxides and $CaCO_3$[a]

Treatment	Rate coefficients $\times 10^3$ (h^{-1})		
	278 K	298 K	318 K
MnO_2 (no coating)	126 ± 13(a)	267 ± 6(a)	533 ± 38(a)
$CaCO_3$ coating (Ca/Mn = 0.08)	39 ± 2(b)	188 ± 15(b)	456 ± 28(a)
$CaCO_3$ coating (Ca/Mn = 0.32)	18 ± 3(b)	39 ± 5(c)	73 ± 10(b)
Al oxide coating (Al/Mn = 0.10)	35 ± 2(c)	257 ± 10(a)	1265 ± 122(c)
Al oxide coating (Al/Mn = 0.40)	11 ± 1(c)	119 ± 5(d)	376 ± 13(a)
Fe oxide coating (Fe/Mn = 0.42)	70 ± 9(d)	318 ± 17(e)	844 ± 22(d)
Fe oxide coating (Fe/Mn = 1.58)	51 ± 6(d)	100 ± 4(f)	172 ± 2(e)

[a] From Oscarson *et al.* (1983a), with permission. For the coated MnO_2, at a given temperature, values followed by (a) are not significantly different ($p < 0.05$) from those of the untreated MnO_2 according to Duncan's multiple range test; within each mineral group ($CaCO_3$, Al oxide, or Fe oxide) of two coating levels at a given temperature, the values followed by the same letter are not significantly different ($p < 0.05$). Values are given as rate constant ± standard error.

can be seen in the reduction of the magnitude of the rate coefficients as the coatings increase. This shows that electron-accepting sites on MnO_2 are partially masked by the oxides and $CaCO_3$. It is known that Al, Fe, and Mn oxides all sorb As, with Al and Fe oxides sorbing more As than MnO_2 (Oscarson *et al.*, 1983a).

In another study, three Mn oxides—birnessite, cryptomelane, and pyrolusite—were investigated as to their ability to deplete As(III) in solution (Oscarson *et al.*, 1983b). Rate coefficients at 298 K were 0.267 and 0.189 h^{-1} for birnessite and cryptomelane, respectively. For pyrolusite, the rate of depletion was much lower, with the rate coefficient being 0.44×10^{-3} h^{-1}. These differences were related to the crystallinity and specific surface of the Mn oxides. Pyrolusite is highly ordered and has a low specific surface of 7.9 m^2 g^{-1}, while birnessite and cryptomelane are poorly crystalline with a high specific surface of 277 and 346 m^2 g^{-1}, respectively (Oscarson *et al.*, 1983b). Oscarson *et al.* (1983b) speculated that As(III) depletion by the Mn oxides was a diffusion-controlled process, since E values ranging from 26.0 ± 0.2 to 32.3 ± 6.7 kJ mol^{-1} were observed.

Cr(III) and Pu(III/IV) Oxidation Kinetics

The oxidation kinetics of Cr(III) and Pu(III/IV) in soils and by Mn oxides has been studied by Amacher and Baker (1982). These two elements have similar behavior in aqueous environments (Bartlett and James,

1979; Rai and Serne, 1977; Amacher and Baker, 1982). Both elements can exist in multiple oxidation states in aqueous environments. Chromium can exist in trivalent and hexavelent states, while Pu can occur in trivalent, quadrivalent, pentavalent, and hexavalent states. Additionally, both elements can exist as cationic or anionic species in aqueous systems. Trivalent Cr exists as the cation Cr^{3+} and its hydrolysis products, or as the anion CrO_2^- at very low concentrations. Hexavalent Cr occurs as the dichromate $Cr_2O_7^{2-}$ or chromate $HCrO_4^-$ or CrO_4^{2-} anions, depending on pH. Plutonium exists in cationic states such as Pu^{3+} and PuO_2^+ and anionic forms such as $PuO_2(CO_3OH^-)$.

Trivalent Cr and Pu(III/IV) cations are immobile in most aqueous and soil environments, since they sorb to soil components. However, Cr(VI) and Pu(VI) anions are quite mobile in soils and aqueous systems, since they are not sorbed by soil constituents to any extent. Consequently, in these forms, they are readily bioavailable (Amacher and Baker, 1982) and are toxic. Hexavalent Cr is suspected of being a human carcinogen. It has been shown that Cr(III) and Pu(III/IV) can be oxidized to Cr(VI) and Pu(VI) by Mn(III/IV) oxides (Cleveland, 1970; Amacher and Baker, 1982).

Kinetics of Cr(III) oxidation in a dilute suspension of a Hagerstown loam soil is shown in Fig. 8.3. The amount of soil used was just enough to oxidize all the added Cr. Also, not all of the Mn oxide was accessible to the Cr and about 50% of the added Cr was sorbed and not available for

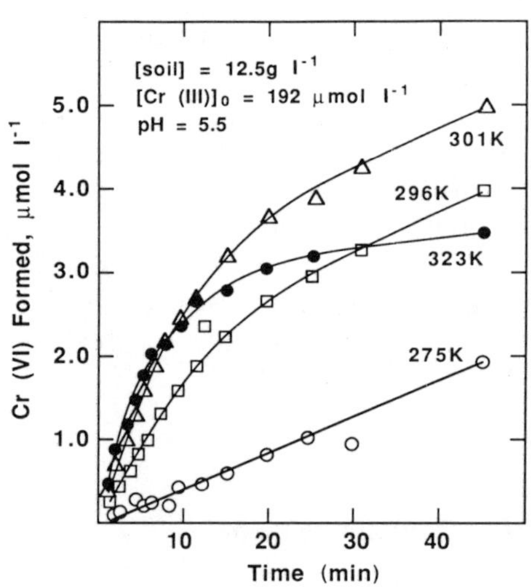

Figure 8.3. Effect of temperature on the kinetics of Cr(III) oxidation in moist Hagerstown silt loam soil. [From Amacher and Baker (1982), with permission.]

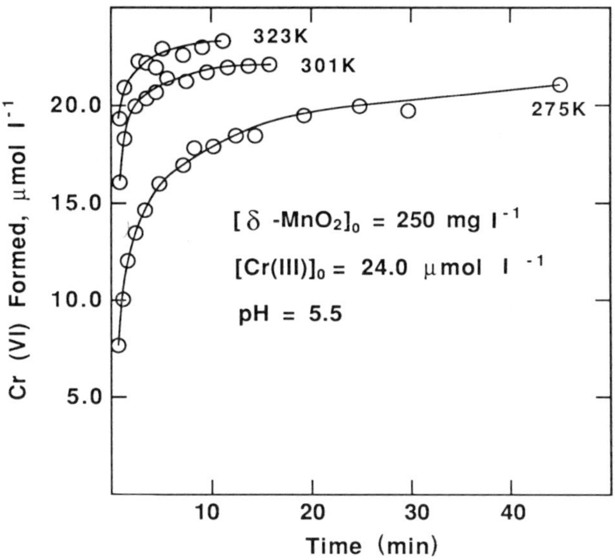

Figure 8.4 Effect of temperature on the kinetics of Cr(III) oxidation by δ-MnO_2 at pH 5.5. [From Amacher and Baker (1982) with permission.]

oxidation. Thus, the actual amounts able to react were different from the theoretical maximum. Most of the oxidation occurred during the first hour. At 274 K, the reaction followed zero-order kinetics. Amacher and Baker (1982) theorized that at low temperature, the oxidation reaction was slowed down considerably so that the Mn oxide surface was "saturated" with Cr undergoing oxidation. At higher temperatures (Fig. 8.3), initially there was a rapid increase in Cr(VI) formed, followed by a decrease in oxidation. However, as Amacher and Baker (1982) correctly note, the study of redox kinetics in soils is different due to sorption and redox reactions occurring simultaneously.

Kinetics of Cr(III) oxidation on Mn(III/IV) oxides showed a trend similar to that observed for soils (Amacher and Baker, 1982). The effect of temperatures and the shape of the kinetic curves are similar (Fig. 8.3 versus Fig. 8.4). Even though a large excess of δ-MnO_2 was used, all of the Cr(III) was not used up (Fig. 8.4). The rapid drop-off in the reaction rate and lack of complete oxidation of Cr(III) was ascribed to part of the reduced Mn(II) not being released to solution as Cr(III) was oxidized. This Mn(II) was probably retained on the oxide surface and prevented fresh Mn(IV) from being available to oxidize Cr(III) (Amacher and Baker, 1982). Morgan and Stumm (1964) found that γ-MnO_2 has a high affinity for its own divalent metal ion.

SUPPLEMENTARY READING

Amacher, M.L., and Baker, D. E. (1982). "Redox Reactions Involving Chromium, Plutonium, and Manganese in Soils, "DOE/DP/OY515.1. Inst. Res. Land and Water Resour. Pennsylvania State University, University Park.

Oscarson, D. W., Huang, P. M., and Hammer, U. T. (1983). Oxidation and sorption of arsenite by manganese dioxide as influenced by surface coatings of iron and aluminum oxides and calcium carbonate. *Water, Air, Soil Pollut.* **20,** 233–244.

Oscarson, D. W., Huang, P. M., Liang, W. K., and Hammer, U. T. (1983). Kinetics of oxidation of arsenite by various manganese oxides. *Soil Sci. Soc. Am. J.* **47,** 644–648.

Stone, A. T. (1986). Adsorption of organic reductants and subsequent electron transfer on metal oxide surfaces. *ACS Symp. Ser,* **323,** 446–461.

Stone, A. T. (1987). Microbial metabolites and the reductive dissolution of manganese oxides: Oxalate and pyruvate. *Geochim. Cosmochim. Acta* **51,** 919–925.

Stone, A. T. (1987). Reductive dissolution of manganese (III/IV) oxides by substitute phenols. *Environ. Sci. Technol.* **21,** 979–988.

9

Kinetic Modeling of Inorganic and Organic Reactions in Soils

INTRODUCTION

A number of kinetically based models have appeared in the literature that describe organic and inorganic reactions in soils. Transport and non-transport models have been used that assume reversible and/or irreversible kinetic reactions.

A number of investigators have modeled solute transport in soils assuming an equilibrium occurs between solution and solid phases. This assumption is often not valid in heterogeneous soil systems, and has been the impetus for the development of a number of nonequilibrium models. Some researchers have assumed that the nonequilibrium is caused by stagnant zones, which result in tortuous diffusional processes between solution and sorbed phases (Rao *et al.*, 1979). Other researchers have attributed the nonequilibrium to kinetic effects.

The use of nonequilibrium kinetic models offers several advantages over equilibrium models. Equilibrium models are often more valid for homogeneous surfaces, and assume that only one type of sorption site is responsible for sorption. Soils are heterogeneous, containing several components (clay minerals, humic substances, oxides, etc.) that exhibit multiple types of sorption sites. Additionally, all of the sorption sites in soils may not be equally accessible. Moreover, there may be sites that exhibit different sorption mechanisms.

Many of the nonequilibrium models that have been used to describe soil chemical reactions take into account the heterogeneous nature of soils. Some of these models have described sorption using a two-site model (Selim *et al.*, 1976b; Cameron and Klute, 1977) characterized by fast and slow binding sites.

One approach to integrating and synthesizing a large body of knowledge on soil elemental transformations and transport is mathematical modeling and computer simulation. In this chapter, aspects of kinetic models dealing with organic and inorganic reactions in soils and on soil constituents will be discussed. For more in-depth reviews, the reader is urged to consult Tanji (1982) for modeling of nitrogen reactions, Berkheiser *et al.* (1980) and Mansell and Selim (1981) for modeling of phosphorus, Iskandar (1981) for wastewater modeling, and Rao and Jessup (1982, 1983) for modeling of pesticide reactions.

MODELING OF INORGANIC REACTIONS

Nitrogen Reactions

A number of kinetically based models to describe nitrogen reactions in soils are presented in Table 9.1. The model of Mehran and Tanji (1974) will be elaborated on since it includes many aspects of nitrogen dynamics in soils.

This model assumes irreversible first-order kinetics for nitrification, denitrification, mineralization, immobilization, and plant uptake. It takes the following form, in which sinks and sources are aggregated:

$$\frac{d(N_c)}{dt} = -\sum_{i=1}^{n} k_i[N_c] + \sum_{j=1}^{m} k_j[N_m] \qquad (9.1)$$

where N_c is the concentration of nitrogen species of interest,
N_m is the concentration of other nitrogen species,
k_i and k_j are respective first-order rate coefficients for $i = 1, \ldots, n$ sink, $j = 1, \ldots, n$ source mechanisms, and t is time.

TABLE 9.1 Dynamic Nitrogen Simulation Models and Subsequent Modification and/or Applications[a]

References	Brief description and comments
Beek and Frissel (1973)	Growth of nitrifier and ammonifer bacteria by Michaelis–Menten kinetics; NH_4^+ oxidation by first-order kinetics with environmental variables; mineralization of proteins, sugars, cellulose, lignin, and living biomass by first-order kinetics; immobilization by first-order kinetics including considerations for microbial biomass and C/N ratio; NH_3 volatilization by diffusion; NH_4^+ clay fixation by equilibrium model.
Mehran and Tanji (1974)	Proposed irreversible first-order kinetics for nitrification, denitrification, mineralization, immobilization, and plant uptake and reversible first-order kinetics for NH_4^+ ion exchange. Model verified with published incubation data.
Shaffer *et al.* (1977)	Dutt *et al.* (1972) model extended to tile-drained croplands and incorporated into a large irrigation return flow model to handle nitrogen as well as dissolved mineral constituents. Model allows user to select the degree of sophistication in simulation or to bypass certain subroutines. Zero-order denitrification and transition-state nitrification added. Model verified for salinity but not nitrogen.
Davidson *et al.* (1978a,b) Davidson and Rao (1978)	Two models developed. Simplified management model: assumes homogeneous profile, displacement of resident soil water and solutes ahead of infiltrating water at field capacity, first-order kinetics for nitrogen transformations (Mehran and Tanji, 1974). Water extraction related to PET and Molz and Remson (1970) model and nitrogen uptake by empirical Michaelis–Menten type of model regulated by water uptake and available NO_3^- and NH_4^+. Refined research model: considers simultaneous transport of water and nitrogen on multilayered soils, first-order kinetics for nitrogen transformation, modification of Molz and Remson extraction model with root growth and root length distribution simulated and constraints on actual transpiration, and plant uptake of nitrogen by Michaelis–Menten kinetics.
Frissel and van Veen (1981)	Revised version of Beek and Frissel (1983) growth of *Nitrosomonas* and *Nitrobacter* by Michaelis–Menten kinetics, including considerations of oxygen levels; denitrification by a physical–biological model involving oxygen diffusion;

(continued)

TABLE 9.1 (*Continued*)

References	Brief description and comments
	mineralization–immobilization considers organic matter grouped into fresh applied organic matter such as animal manures, straw and waste water, and resistant biomass residues; NH_3 volatilization by physicochemical model including dissociation of NH_4OH; NH_4^+ clay fixation by reversible first-order kinetics.
Reddy *et al.* (1979a,b)	Donigan and Crawford (1976) model extended to handle animal waste loadings. Modified mineralization simulation with considerations for potentially mineralized nitrogen and short-term rate kinetics. Added ammonia volatilization from animal wastes that is dependent on temperature, air flow rate, and CEC. Mineralization and volatilization validated with laboratory data.
Tanji and Mehran (1979)	Same as Mehran and Tanji (1974) except nitrogen uptake by variable NO_3^- and NH_4^+ absorption coefficients.
Tillotson *et al.* (1980)	Nitrification and urea hydrolysis by first-order kinetics; NH_3 volatilization by first-order kinetics from $(NH_4)_2CO_3$ formed from urea hydrolysis; NH_4^+ sorption by linear partition model; NH_4^+ and NO_3^- plant uptake involving diffusion to roots.

[a] From Tanji (1982), with permission.

For example, the rate of change in solution NH_4^+ concentration with time is given by (Fig. 9.1)

$$d(\mathrm{NH_4})_s/dt = -(k_1 + k_{se} + k_4 + k_6)(\mathrm{NH_4})_s + k_{es}(\mathrm{NH_4})_e + k_6(\mathrm{Org}\ N)_i \qquad (9.2)$$

with terms as defined in Fig. 9.1.

Mehran and Tanji (1974) mentioned several problems with the above approach, including (1) assuming first-order kinetics, (2) empirical fitting of rate coefficients, (3) rate coefficients that are constant, implying that concentrations of substrates are not limiting, and (4) no consideration for environmental factors such as temperature, soil-water content, and aeration.

Tanji and Mehran (1979) described NH_4^+ and NO_3^- plant uptake by a variable absorption scheme such that

$$\text{Uptake} = \lambda CS(z, \mathrm{t})/\theta \qquad (9.3)$$

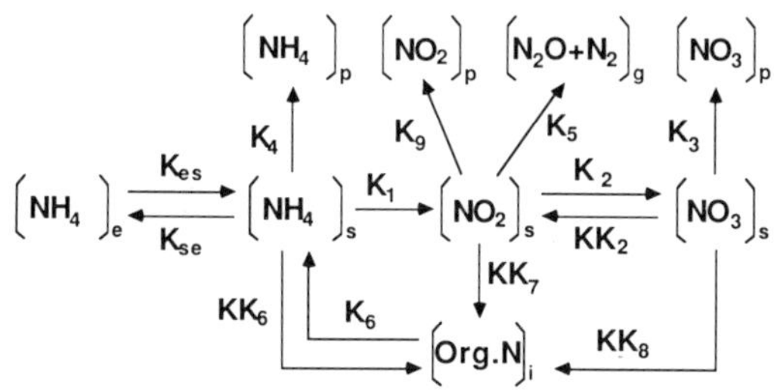

Figure 9.1. Possible rate transformations of soil nitrogen. Terms *k* and *kk* denote rate coefficients; e, s, p, i, and g refer to exchangeable, solution, plant, immobilized, and gaseous phases, respectively. [From Mehran and Tanji (1974), with permission.]

where λ is the NH_4^+ absorption coefficient, which can be set to unity if uptake is assumed proportional to root water extraction, C is the concentration of solution NH_4^+, S is the rate of extraction, and θ is the soil water content. A similar equation was used for the absorption of NO_3^- by plant roots.

Davidson *et al.* (1978a) also used first-order kinetics for nitrogen transformations, but they considered that some of the transformation rate coefficients were dependent on several factors including environmental ones The rate for nitrification was empirically adjusted for water suction. Overall, most of the nitrogen models assume first-order kinetics. Some of them also consider the effects of temperature on rate coefficients.

Phosphorus Reactions

Several kinetic models have appeared to describe phosphorus reactions in soils. Enfield (1978) classified models for estimating phosphorus concentrations in percolate waters derived from soil that had been treated with wastewater into three categories: (1) empirical models that are not based on known theory; (2) two-phase kinetic models that assume a solution phase and some adsorbed phase; and (3) multiphase models, which include solution, adsorbed, or precipitated phases. Mansell and Selim (1981) classified models as shown in Table 9.2. The reader is urged to consult this reference for a complete discussion of the phosphorus kinetic models. For the purpose of this discussion, attention will be given to models that assume reversible phosphorus removal from solution, which can occur simultaneously by equilibrium and nonequilibrium reactions, and mechanistic multiphase models for reactions and transport of phosphorus applied to soils.

TABLE 9.2 Summary of Mathematical Models for Predicting Phosphorus Reactions in Soils

Type and model	Reference
Mathematical models that assume chemical nonequilibrium	
Transport models that assume reversible kinetic reactions for applied phosphorus	Selim *et al.* (1976b)
Transport models that assume irreversible kinetic reactions for applied phosphorus	Cho *et al.* (1970) Panda *et al.* (1978)
Transport models that assume both reversible and irreversible reactions for applied phosphorus	Overman *et al.* (1976) Mansell *et al.* (1977)
Nontransport sorption models that assume both reversible and irreversible kinetic reactions for applied phosphorus	Overman and Chu (1977a) Overman and Chu (1977b) Overman and Chu (1977c)
Mathematical models that assume reversible phosphorus removal from solution to occur simultaneously by equilibrium and nonequilibrium reactions	
Transport models that assume two types of phosphorus sorption sites	Selim *et al.* (1976b) DeCamargo *et al.* (1979)
Nontransport models that assume two types of phosphorus sorption sites	Fiskell *et al.* (1979)
Mechanistic multiphase models for reactions and transport of phosphorus applied to soils	Mansell *et al.* (1977a)

Transport Models That Assume Two Types of Phosphorus Sorption Sites. Mansell and Selim (1981) assumed applied soil phosphorus can react reversibly with two broad kinds of sorption sites—those exhibiting high rates, and those showing low rates. Such a model can be represented as

$$\underset{\text{sorbed P}}{\rho_B q_{II}} \underset{}{\overset{\text{slow}}{\rightleftharpoons}} \underset{\text{solution P}}{\theta C} \overset{\text{rapid}}{\rightleftharpoons} \underset{\text{sorbed P}}{\rho_B q_I} \tag{9.4}$$

where q_I and q_{II} are the amounts of phosphorus (P) sorbed on the sites showing high and low rates, respectively, ρ_B is the bulk density, and θ is the volumetric water content. The total amount of sorbed phosphorus (q_T) would be $q_T = q_I + q_{II}$.

Selim *et al.* (1976b) developed a simplified two-site model to simulate sorption–desorption of reactive solutes applied to soil undergoing steady water flow. The sorption sites were assumed to support either instantaneous (equilibrium sites) or slow (kinetic sites) first-order reactions. As pore-water velocity increased, the residence time of the solute decreased and less time was allowed for kinetic sorption sites to interact (Selim *et al.*, 1976b). The sorption–desorption process was dominated by the equilibrium sites when rapid velocities were present and by both types of sites when slow velocities were operational.

De Camargo *et al.* (1979) used the model of Selim *et al.* (1976b) to study sorption–desorption reactions on soil using miscible displacement and batch techniques. The model adequately described breakthrough curves for a range of soil aggregate sizes and pore water velocities. Additionally, De Camargo *et al.* (1979) felt that it could differentiate between instantaneous and time-dependent reactions controlling phosphorus sorption and desorption phenomena under steady soil-water flow conditions.

Nontransport Model That Assumes Two Types of Phosphorus Sorption Sites. Fiskell *et al.* (1979) studied phosphorus sorption on soils and found that the two-side model of Selim *et al.* (1976b) described their data much better than a one-site nonlinear kinetic model.

Mechanistic Multiphase Model for Reactions and Transport of Phosphorus Applied to Soils. Mansell *et al.* (1977a) presented a mechanistic model for describing transformations and transport of applied phosphorus during water flow through soils. Phosphorus transformations were governed by reaction kinetics, whereas the convective–dispersive theory for mass transport was used to describe P transport in soil. Six of the kinetic reactions—adsorption, desorption, mobilization, immobilization, precipitation, and dissolution—were considered to control phosphorus transformations between solution, adsorbed, immobilized (chemisorbed), and precipitated phases. This mechanistic multistep model is shown in Fig. 9.2.

For no-water-flow conditions—that is, when net movement of the soil solution is considered negligible—mathematical equations that simultaneously describe rates of phosphorus transfer between A, B, C, and D phases may be given as follows (Mansell *et al.*, 1977a):

$$\partial(\theta A)/\partial t = -\theta(k_a A^n + k_3 A) + \rho_B(k_d B + k_4 D) \tag{9.5}$$

$$\partial(\rho_B B)/\partial t = k_a \theta A^n - (k_d + k_1)\rho_B B + k_2 \rho_B C \tag{9.6}$$

$$\partial(\rho_B C)/\partial t = k_1 \rho_B B - k_2 \rho_B C \tag{9.7}$$

$$\partial(\rho_B D)/\partial t = k_3 \theta A - k_4 \rho_B D \tag{9.8}$$

where A is the concentration of phosphorus in solution; B, C, and D are the amounts of phosphorus per gram soil for adsorbed, immobilized, and precipitated phases, respectively; n is a constant representing the order of the adsorption process; k_a, k_d, k_1, k_2, k_3, and k_4 are rate coefficients for adsorption, desorption, immobilization, mobilization, precipitation, and dissolution, respectively; and θ, t, and ρ_B were previously defined.

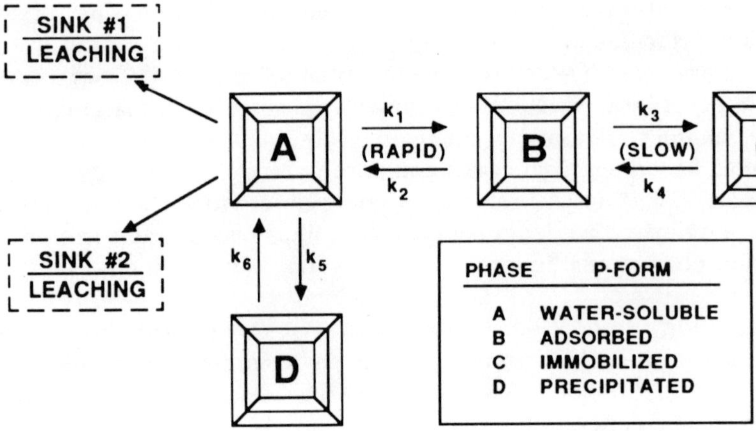

Figure 9.2. A schematic representation of six reversible-kinetic reactions that are assumed to control the transfer of applied phosphorus (P) between solution, adsorbed, immobilized (chemisorbed), and precipitated phases within the soil. Sinks are shown for irreversible removal of phosphorus from the soil solution by plant uptake and by leaching. [From Mansell *et al.* (1977a), with permission.]

Under no-flow conditions where sufficient time had passed to establish chemical equilibrium, the right-hand side of Eqs. (9.5)–(9.8) would equal zero and could be solved simultaneously to provide a relationship between the total sorbed quantity q of phosphorus and the concentration A of water-soluble phosphorus. For the linear case where the order of the reaction $n = 1$, the relationship is

$$q = D_g A = \frac{\theta}{\rho_B}\left[\frac{k_a}{k_d}\left(1 + \frac{k_1}{k_2}\right) + \frac{k_3}{k_4}\right]A \tag{9.9}$$

where D_g is the distribution coefficient as obtained (Mansell *et al.*, 1977b) from solute adsorption isotherms. However, for the nonlinear case ($n \neq 1$) as normally observed for phosphorus the relationship becomes

$$q = (D_{g1}A^{1-n} + D_{g2})A^n \tag{9.10}$$

where the distribution coefficient D_g is no longer a constant but is dependent on concentration. The constants D_{g1} and D_{g2} are defined as

$$D_{g1} = \left(\frac{\theta}{\rho_B}\right)\frac{k_3}{k_4} \tag{9.11}$$

$$D_{g2} = \left(\frac{\theta}{\rho_B}\right)\frac{k_a}{k_d}\left(1 + \frac{k_1}{k_2}\right) \tag{9.12}$$

Mansell *et al.* (1977b) presented simulated results for a range of rate coefficients (reflecting a range of soil characteristics), that is, phosphorus sorption capacities, for the kinetic transformations during steady water flow initially devoid of phosphorus.

Potassium Reactions

Selim *et al.* (1976a) proposed a mathematical model for potassium reactions and transport in soils. Kinetic reactions were assumed to govern the transformation between solution, exchangeable, nonexchangeable (secondary minerals), and primary mineral phases of potassium shown in Fig. 9.3.

Mathematical derivations of potassium transport and transformation processes may be formulated as follows. The following new terms can be defined: C, concentration of potassium in solution phase; S_1, amount of potassium in exchangeable phase; S_2, amount of potassium in nonexchangeable phase; S_3, amount of potassium in primary mineral phase; u_{PW}, pore water velocity; D_c, dispersion coefficient; and d, depth or distance below soil surface.

The transport and transformation processes of potassium in the solution phase may be expressed (Selim *et al.*, 1976a) as

$$\frac{\partial C}{\partial t} = D_c \frac{\partial^2 C}{\partial d^2} - \frac{u_{PW}\, \partial C}{\partial d} - k_a C^n + (\rho_B/\theta) k_d S_1 \tag{9.13}$$

The first two terms of the right side of Eq. (9.13) account for transport. These two terms are referred to as the dispersion and mass flow terms, respectively. The third and fourth terms represent the adsorption (forward) and desorption (backward) reactions between the exchangeable and solution phases. Selim *et al.* (1976a) assumed the backward reaction was a

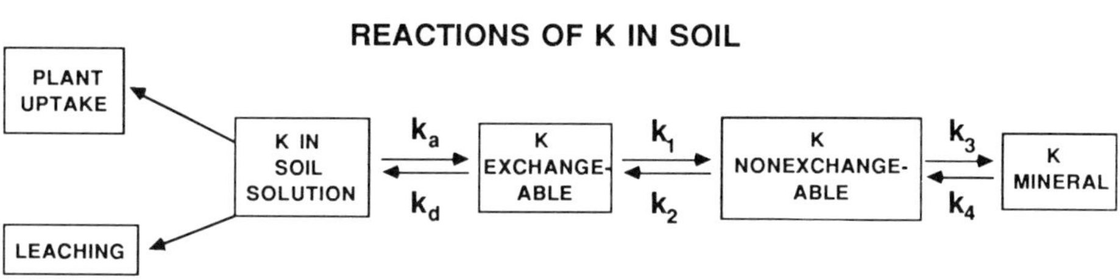

Figure 9.3. Schematic representation of the reactions of potassium (K) in solution, exchangeable, nonexchangeable (complex secondary minerals), and primary mineral phases in soil. [From Selim *et al.* (1976a), with permission.]

kinetic type that was first-order, whereas the forward reaction was n^{th}-order and in the authors' work was assumed to be less than unity.

The transformation of the exchangeable phase (S_1) may be written as

$$\partial S_1/\partial t = (\theta/\rho_B)k_a C^n + k_2 S_2 - (k_1 k_d)S_1 \tag{9.14}$$

Similarly, the transformations of the nonexchangeable phase (S_2) and of the mineral phase (S_3) could be expressed, respectively, as

$$\partial S_2/\partial t = k_1 S_1 + k_4 S_3 - (k_2 + k_3)S_2 \tag{9.15}$$

$$\partial S_3/\partial t = k_3 S_2 - k_4 S_3 \tag{9.16}$$

In Eqs. (9.14)–(9.16), the reactions between exchangeable and nonexchangeable and those between nonexchangeable and primary mineral phases were assumed to be first-order kinetic reactions.

The differential Eqs. (9.13)–(9.16) that described transport and reactions of potassium phases may be regarded as a representation of the model shown in Fig. 9.3. The initial and boundary conditions chosen were (Selim *et al.*, 1976a)

$$C = 0 \qquad t = 0 \qquad 0 < d < l \tag{9.17}$$

$$S_1 = S_2 = S_3 = 0 \qquad t = 0 \qquad 0 < d < l \tag{9.18}$$

$$u_{PW} C - D_c \frac{\partial C}{\partial d} = u_{PW} C_0 \qquad t < t_1 \qquad d = 0 \tag{9.19}$$

$$u_{PW} C - D_c \frac{\partial C}{\partial d} = 0 \qquad t > t_1 \qquad d = 0 \tag{9.20}$$

$$\frac{\partial C}{\partial d} = 0 \qquad t > 0 \qquad d = l \tag{9.21}$$

which describe potassium transport and transformation in a soil column of length l, initially ($t = 0$) devoid of potassium. A solution having soluble potassium concentration C_0 was applied at the soil surface for a length of time t_1 and was then followed by potassium-free solution (Selim *et al.*, 1976a). A steady pore water velocity u_{PW} throughout the soil was assumed at all times.

The value of the parameter n in Eqs. (9.13)–(9.14) is different from unity, and these differential equations are nonlinear and cannot be solved analytically. Therefore, Eqs. (9.13)–(9.16) subject to the conditions of Eqs. (9.17)–(9.21) were solved using numerical analysis techniques. Selim *et al.* (1976a) used the explicit–implicit finite-difference approximations as the method of solution. This was successfully used by Selim *et al.* (1975) for steady water flow conditions and by Selim *et al.* (1976a) for transient water conditions. Finite-difference approximations provide distributions of the various potassium phases (C, S_1, S_2, and S_3), at incremental distances d in the soil and at a discrete time steps t. A mass balance was maintained as a check on the numerical results (Selim *et al.*, 1976a).

Aluminum Reactions

Jardine *et al.* (1985b) employed a two-site nonequilibrium transport model to study Al sorption kinetics on kaolinite. They used the transport model of Selim *et al.* (1976b) and Cameron and Klute (1977). Based on the above model, Jardine *et al.* (1985a) concluded that there were at least two mechanisms for Al adsorption on Ca-kaolinite. It appeared that there were equilibrium (type-1) reactions on kaolinite that involved instantaneous Ca–Al exchange and rate-limited reaction sites (type-2) involving Al polymerization on kaolinite. The experimental breakthrough curves (BTC) conformed well to the two-site model.

MODELING OF SOIL–PESTICIDE INTERACTIONS

A number of kinetically based models have been used to study soil–pesticide reactions. In many cases, sorption of pesticides has been treated as a rapid-equilibrium, single-valued, reversible process. Some of these models are briefly outlined below.

Van Genuchten *et al.* (1974) assumed that the amount of pesticide sorbed with time followed reversible nonlinear kinetics such that

$$\frac{\partial q}{\partial t} = k_{-1}\left[\left(\frac{k_1 \theta}{k_{-1}\rho_B}\right)C^N - q\right] \tag{9.22}$$

where k_1 and k_{-1} are forward and backward rate coefficients (h^{-1}), θ is soil water content ($m^3\ m^{-3}$), ρ_B is bulk density (gm^{-3}), t is time (h), C is solution-phase pesticide concentration ($mg\ l^{-1}$), and q is amount of pesticide sorbed ($mg\ kg^{-1}$ soil). At equilibrium $\partial q/\partial t = 0$ and Eq. (9.22) reduces to

$$q = k_F C^N \tag{9.23}$$

where k_F and N are coefficients that depend on the soil–pesticide system and the Freundlich coefficient k_F is equal to $\theta k_1/\rho_B k_{-1}$.

The sorption kinetics of 2,4-D on illite, kaolinite, and montmorillonite was modeled by Haque *et al.* (1968) using

$$d/dt = (1 - \phi) = k_q(1 - \phi) \tag{9.24}$$

where $\phi = q_t/q_{eq}$ and q_t and q_{eq} are the amounts of pesticide sorbed at time t and at equilibrium (expressed as mg kg^{-1}) and k_q is the sorption rate coefficient (h^{-1}).

The rate of sorption in Eq. (9.24) is proportional to the distance from equilibrium. As noted earlier, in deriving Eq. (9.24), which is based on the generalized equilibrium theory of Fava and Eyring (1956), it is assumed that the reverse reaction and desorption rate are small enough to be neglected. Haque *et al.* (1968) satisfied this assumption by using large amounts of each of the clays (5–15 g) and low 2,4-D solution concentrations (1.3 mg l^{-1}).

Lindstrom *et al.* (1970) proposed a model that is an extension of the model of Fava and Eyring (1956), in which they added a "sticking probability" for the sorbate on the sorbent surface, which is permitted to vary with the degree of surface coverage. Their model allowed for sorption and desorption energies to change with surface coverage. The model of Lindstrom *et al.* (1970) can be stated as

$$\partial q/\partial t = [k_{-1} \exp(\beta q)]\left[\left(\frac{k_1\,\theta}{k_{-1}\,\rho_B}\right) C \exp(-2\beta q) - q\right] \qquad (9.25)$$

where β is similar to the surface stress coefficient described by Fava and Eyring (1956) and other terms are defined as given in Eq. (9.22). At equilibrium, $\partial q/\partial t = 0$, and Eq. (9.25) reduces to

$$q = (\theta k_1/\rho_B k_{-1})[C \exp(-2\beta q)] \qquad (9.26)$$

This can be rearranged as

$$\ln(q/C) = \ln(\theta k_1/\rho_B k_{-1}) - 2\beta q \qquad (9.27)$$

The parameters β and $(\theta k_1/\rho_B k_{-1})$ can be obtained from the slope and intercept, respectively, of plots of $\ln(q/C)$ versus various equilibrium solution concentrations.

An example of a two-site model for pesticide desorption kinetics from soils was presented by McCall and Agin (1985). Using a reversible first-order equation to describe picloram desorption from soil, the authors plotted $\ln(C_B - C_{Beq})$ versus t, where C_B is the bound form of picloram and C_{Beq} is the bound form at equilibrium, in Fig. 9.4. It is obvious that desorption conforms to a two-step process where a fast step occurs for about a 5-h period, and then a slow process as shown from the linear portion of Fig. 9.4 occurs from 5 to 300 h. McCall and Agin (1985) used the following two-step model to describe the data shown in Fig. 9.4:

$$C_{B1} \xrightarrow{k_1} C_F \underset{k_{-2}}{\overset{k_2}{\rightleftharpoons}} C_{B2} \qquad (9.28)$$

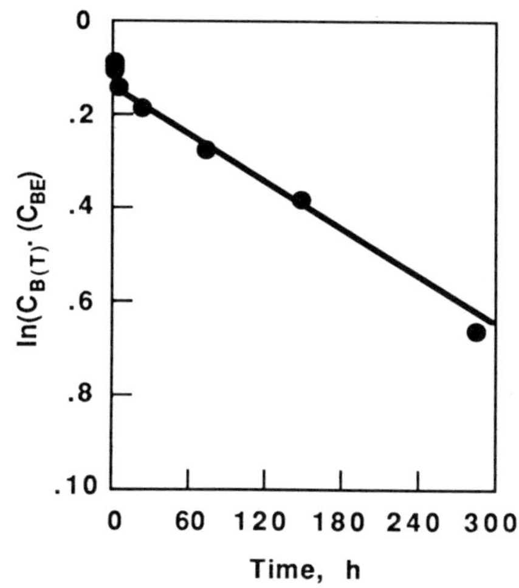

Figure 9.4. Treatment of picloram desorption data as a simple first-order reversible process for sample incubated 100 days in Kawkawlin soil. Plotted line is computed from model. [From McCall and Agin (1985), with permission.]

where C_{B1} is the picloram bound at fast desorbing sites, C_{B2} is the picloram retained at slow desorbing sites, C_F is the free form of picloram, k_1 is the fast desorption rate coefficient, k_2 is the slow desorption rate coefficient, k_{-2} is the reversible slope sorption rate coefficient, and C_{BT} is $C_{B1} + C_{B2}$. Thus,

$$d(C_{B1})/dt = -k_1(C_{B1}) \tag{9.29}$$

$$d(C_F)/dt = k_1/(C_{B1}) + k_2(C_{B2}) - k_{-2}(C_F) \tag{9.30}$$

$$d(C_{B2})/dt = k_2(C_{B2}) - k_{-2}(C_F) \tag{9.31}$$

Equations (9.29)–(9.31) were solved numerically with parameter estimation routines written for use with an IBM Continuous System Modeling Program (CSMP III). Relative least-squares minimization was performed.

In the model of McCall and Agin (1985), picloram desorption occurs from two different soil sites. Picloram could be released rapidly from one site (B1) characterized by rate constant k_1 and slowly and reversibly from the other site by rate constants k_2 and k_{-2}. The fast steps may be reversible; however, in the experiments of McCall and Agin (1985) the backward rate coefficient was small since this step could be described with one rate coefficient. The plotted line in Fig. 9.4 is computed from the previous model. The model also fits the data well at early desorption times (Fig. 9.5). McCall and Agin (1985) found that k_1 values were approximately two orders of magnitude higher than k_2 and k_{-2}. Other parameters

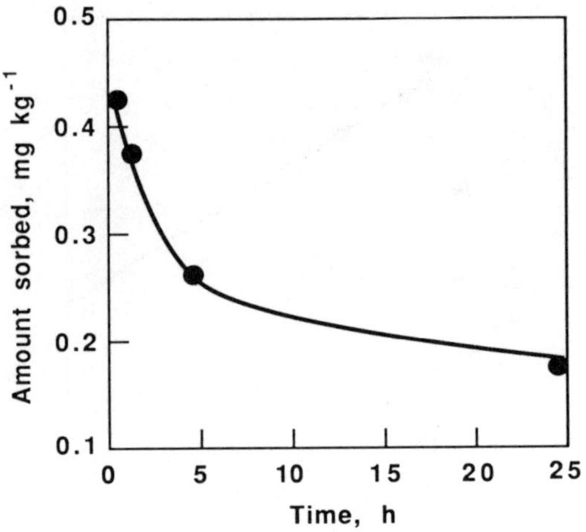

Figure 9.5. Computer fit of picloram desorption data from 100-day sample with Kawkawlin soil over initial portion of the desorption curve. [From McCall and Agin (1985), with permission.]

one can calculate from the model are the concentrations of pesticide sorbed at each site initially (C_{B10} and C_{B20}) (time = 0) and the concentration bound at equilibrium (C_{Beq}).

McCall and Agin (1985) speculated that the rapid sorption–desorption of picloram on soils was associated with external sites of organic matter, whereas the slower kinetics could be ascribed to particle diffusion (PD) in organic or inorganic soil components. The amount of picloram associated with the external sites remained nearly constant with time, while that associated with internal sites increased greatly over time.

MODELING OF ORGANIC POLLUTANTS IN SOILS

Wu and Gschwend (1986) reviewed and evaluated several kinetic models to investigate sorption kinetics of hydrophobic organic substances on sediments and soils. They evaluated a first-order model (one-box) where the reaction is evaluated with one rate coefficient (k) as well as a two-site model (two-box) whereby there are two classes of sorbing sites, two chemical reactions in series, or a sorbent with easily accessible sites and difficultly accessible sites. Unfortunately, the latter model has three independent fitting parameters: k_1, the exchange rate from the solution to the first (accessible sites) box; k_2, the exchange rate from the first box to the second (difficultly accessible sites) box; and X_1, the fraction of total sorbing capacity in the first box. Both of these models have problems. The first-order model does not always fit experimental data well, and the parameters from the two-box model are difficult to obtain and evaluate. Accordingly, Wu and Gschwend (1986) developed a radial diffusive penetration model modified by a retardation factor considering microscale partitioning of the sorbate between intraggregate pore fluids and the solids making up the aggregate grains. This model assumes that the sediment and soil particles are aggregates of fine mineral grains and organic matter and that the organics diffuse through the pore fluids between the interstitial regions of the aggregates. Also, their penetration is retarded by microscale partitioning of the organics between essentially mobile (dissolved in particle pore fluids) and immobile (in/on particle solids) states of the chemical.

The time rate of change of sorbed compound per unit volume can be expressed (Crank, 1976) as

$$\frac{\partial S(r)}{\partial t} = D_{\mathrm{m}} p_{\mathrm{s}} \left[\frac{\partial^2 C'(r)}{\partial r^2} + \frac{2}{r} \frac{\partial C'(r)}{\partial r} \right] \tag{9.32}$$

where $S(r)$ is the local total volumetric concentration in the porous sorbent (mol cm^{-3}), $C'(r)$ is the compound concentration free in the pore fluid and changing with radial distance r (mol cm^{-3}), p_{s} is the porosity of the sorbent (cm^3 of fluid cm^{-3} total), and D_{m} is the pore fluid diffusivity of the sorbate (cm^2 s^{-1}). By definition,

$$S(r) = (1 - p_{\mathrm{s}})_{\rho_{\mathrm{s}}} S'(r) + p_{\mathrm{s}} C'(r) \tag{9.33}$$

where $S'(r)$ is the concentration of the immobile state (mol g^{-1}) and ρ_{s} is the specific gravity of the sorbent (g cm^{-3}). If the pore fluid concentration and the solid-bound concentration are locally in equilibrium, a sorption isotherm relating these states applies, such that

$$S'(r) = K_{\mathrm{p}} C'(r)$$

where K_{p} is the equilibrium partition coefficient [(mol g^{-1}/(mol cm^{-3})]. The isotherm can be used to restate the particle diffusion kinetics in terms of S only:

$$S(r) = (1 - p_{\mathrm{s}})\rho_{\mathrm{s}} K_{\mathrm{p}} C'(r) + p_{\mathrm{s}} C'(r) = [(1 - p_{\mathrm{s}})\rho_{\mathrm{s}} K_{\mathrm{p}} + p_{\mathrm{s}})]C(r) \tag{9.34}$$

$$\begin{aligned} \frac{\partial S(r)}{\partial t} &= \frac{D_{\mathrm{m}} p_{\mathrm{s}}}{[(1 - p_s)\rho_{\mathrm{s}} K_{\mathrm{p}} + p_{\mathrm{s}}]} \left[\frac{\partial^2 S(r)}{\partial r^2} + \frac{2}{r} \frac{\partial S(r)}{\partial r} \right] \\ &= D'_{\mathrm{eff}} \left[\frac{\partial^2 S(r)}{\partial r^2} + \frac{2}{r} \frac{\partial S(r)}{\partial r} \right] \end{aligned} \tag{9.35}$$

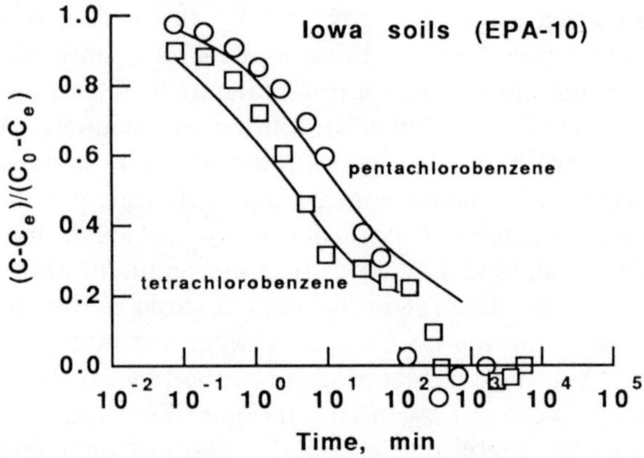

Figure 9.6 Experimental and model-fitting results for tetrachlorobenzene and pentachlorobenzene sorption kinetics on Iowa soils using the retarded/radial diffusion model of Wu and Gschwend (1986): C_s is the dissolved concentration in the bulk solution, and C_0 and C_{eq} are the concentrations initially and at equilibrium, respectively. [From Wu and Gschwend (1986), with permission.]

where D'_{eff} is the effective particle diffusivity ($cm^2\ s^{-1}$). When K_p is large, as is the case for hydrophobic compounds, D'_{eff} is

$$D'_{eff} = \frac{D_m p_s}{(1 - p_s)\rho_s K_p} \tag{9.36}$$

With the above model, Wu and Gschwend (1986) also assumed that the entire surface area is available for mass flux and the path length of diffusive transfer is half the particle diameter. The authors introduced a correction factor $f(p_s, t)$ for D_{eff} for natural silts, which is a function of intraaggregate porosity or tortuosity (tor), that is,

$$\frac{\partial S(r)}{\partial \text{tor}} = D_{eff}\left[\frac{\partial^2 S(r)}{\partial r^2} + \frac{2}{r}\frac{\partial S(r)}{\partial r}\right] \tag{9.37}$$

where $D_{eff} = D'_{eff} f(p_s, t)$. Details of model simulation are given in Wu and Gschwend (1986).

The authors used all three models to study kinetics of sorption of 1,4-dichlorobenzene, 1,2,4-trichlorobenzene, 1,2,3,4-tetrachlorobenzene, and pentachlorobenzene on soils and sediments using a batch technique. The one-box model did not fit the data well, while the two-box model described the data better. However, there are three fitting parameters (k_1, k_2, and x_1) in this model, which could explain the better conformity of the data to the model. Another disadvantage of this model is that it is not easy to relate these three parameters to known properties of the colloidal material. Wu and Gschwend (1986), found, for example, that two totally different sets of parameters for tetrachlorobenzene sorption on a Charles River sediment were obtained if different sediment mean aggregate sizes were used. This indicates that the parameters would have to be experimentally estimated for each type of sediment, which is not practical. However, the authors found that sorption of organic pollutants on the sediments was described very well with the retarded/radial diffusion model by adjusting only the effective intraparticle diffusivity parameter (Fig. 9.6). An additional benefit of this model is that the D_{eff} values can be estimated *a priori* by correlation with chemical and sediment or colloidal properties.

SUPPLEMENTARY READING

Iskandar, I. K., ed. (1981). "Modeling Wastewater Renovation." Wiley, New York.

Mansell, R. S., and Selim, H. M. (1981). Mathematical models for predicting reactions and transport of phosphorus applied to soils. "*In* Modeling Wastewater Renovation" (I. K. Iskander, ed.), pp. 600–640. Wiley, New York.

Mehran, M., and Tanji, K. K. (1974). Computer modeling of nitrogen transformations in soils. *J. Environ. Qual.* **3,** 391–395.

Rao, P. S. C., and Jessup, R. E. (1982). Development and verification of simulation models for describing pesticide dynamics in soils. *Ecol. Modell.* **16,** 67–75.

Selim, H .M., Mansell, R. S., and Zelazny, L. W. (1976). Modeling reactions and transport of potassium in soils. *Soil Sci.* **122,** 77–84.

Tanji, K. K. (1982). Modeling of the nitrogen cycle. *In* "Nitrogen in Agricultural Soils" (F. J. Stevenson, ed.), pp. 721–72. Am. Soc. Agron. Madison, Wisconsin.

Wu, S., and Gschwend, P. M. (1981). Sorption kinetics of hydrophobic organic compounds to natural sediments and soils. *Environ. Sci. Technol.* **20,** 717–725.

Bibliography

Aagaard, P., and Helgeson, H. C. (1982). Thermodynamic and kinetic constraints on reaction rates among minerals and aqueous solutions. I. Theoretical considerations. *Am. J. Soc.* **282,** 237–285.

Aharoni, C., and Ungarish, M. (1976). Kinetics of activated chemisorption. I. The non-Elovichian part of the isotherm. *J. Chem. Soc., Faraday Trans.* **72,** 400–408.

Aharoni, C., and Ungarish, M. (1977). Kinetics of activated chemisorption. II. Theoretical models. *J. Chem. Soc., Faraday Trans.* **73,** 456–464.

Akratanakul, S., Boersma, L., and Klock, G. O. (1983). Sorption processes in soils as influenced by pore water velocity. 2. Experimental results. *Soil Sci.* **135,** 331–341.

Amacher, M. L., and Baker, D. E. (1982) "Redox Reactions Involving Chromium, Plutonium, and Manganese in Soils," DOE/DP/04515-1. Pennsylvania State University, University Park.

Amer, F., Bouldin, D. R., and Black, C. A. (1955). Characterization of soil phosphorus by anion exchange resin adsorption and ^{32}P equilibration. *Plant Soil* **6,** 391–394.

Amrhein, C., Jurinak, J. J., and Moore, W. M. (1985). Kinetics of calcite dissolution as affected by carbon dioxide partial pressure. *Soil Sci. Soc. Am. J.* **49,** 1393–1398.

Andreae, M. O., and Klumpp, D. (1979). Biosynthesis and release of organo-arsenic compounds by marine algae. *Environ. Sci. Technol.* **13,** 738–741.

Ashida, M., Sasaki, M., Kan, H., Yasunaga, T., Hachlya, K., and Inoue, T. (1978). Kinetics of proton adsorption-desorption at TiO_2–H_2O interface by means of pressure jump technique. *J. Colloid Interface Sci.* **678,** 219–225.

Astumian, R. D., Sasaki, M., Yasunaga, T., and Schelly, Z. A. (1981). Proton adsorption-desorption kinetics on iron oxides in aqueous suspensions, using the pressure-jump method. *J. Phys. Chem.* **85,** 3832–3835.

Atkinson, R. J., Hingston, F. J., Posner, A. M., and Quirk, J. P. (1970). Elovich equation for the kinetics of isotope exchange reactions at solid-liquid interfaces. *Nature* (*London*) **226,** 148–149.

Bailey, G. W., and White, J. L. (1970). Factors influencing adsorption, desorption and movement of pesticides in soil. *Residue Rev.* **32,** 29–92.

Bajpai, R. K., Gupta, A. K., and Rao, M. G. (1974). Single particle studies of binary and ternary cation exchange kinetics. *AIChE J.* **20,** 989–995.

Barrow, N. J. (1979). The effects of temperature on the reactions between inorganic phosphate and soil. *J. Soil Sci.* **30,** 271–279.

Barrow, N. J. (1983). A discussion of the methods for measuring the rate of reaction between soil and phosphate. *Fert. Res.* **40,** 51–59.

Barrow, N. J., and Shaw, T. C. (1975). The slow reactions between soil and anions. II. Effect of time and temperature on the decrease in phosphate concentration in the soil solution. *Soil Sci.* **119,** 167–177.

Barrow, N. J., and Shaw, T. C. (1979). Effects of solution: Soil ratio and vigour of shaking on the rate of phosphate adsorption by soil. *J. Soil Sci.* **30,** 67–76.

Bartlett, R. J., and James, B. (1979). Behavior of chromium in soils. III. Oxidation. *J. Environ. Qual.* **8,** 31–35.

Beek, J., and Frissel, M. J. (1973). "Simulation of Nitrogen Behavior in Soil." Centre for Agricultural Publishing and Documentation, Wageningen, The Netherlands.

Berkheiser, V. E., Street, J. J., Rao, P. S. C., and Yuan, T. L. (1980). Partitioning of inorganic orthophosphate in soil-water systems. *CRC Crit. Rev. Environ. Control* **10,** 179–224.

Bernasconi, C. F. (1976). "Relaxation Kinetics." Academic Press, New York.

Bernasconi, C. F., ed. (1986). "Investigations of Rates and Mechanisms of Reactions," 4th ed. Wiley, New York.

Berner, R. A. (1978). Rate control of mineral dissolution under earth surface conditions. *Am. J. Sci.* **278,** 1235–1252.

Berner, R. A. (1980). "Early Diagenesis: A Theoretical Approach." Princeton University Press, Princeton, New Jersey.

Berner, R. A. (1983). Kinetics of weathering and diagenesis. *Mineral.* **8,** 111–134.

Berner, R. A., and Holdren, G. R., Jr. (1977). Mechanism of feldspar weathering: Some observational evidence. *Geology* **5,** 369–372.

Berner, R. A., and Holdren, G. R., Jr. (1979). Mechanism of feldspar weathering. II. Observations of feldspars from soils. *Geochim. Cosmochim. Acta* **43,** 1173–1186.

Berner, R. A., Holdren, G. R., Jr., and Schott, J. (1985). Surface layers on dissolving silicates. (Comments on "Study of the weathering of albite at room temperature and pressure with a fluidized bed rector" by L. Chou and R. Wollast.) *Geochim. Cosmochim. Acta* **49,** 1657–1658.

Best, J. A., Weber, J. B., and Weed, S. B. (1972). Competitive adsorption of diquat^{2+}, paraquat^{2+}, and Ca^{2+} on organic matter and exchange resins. *Soil Sci.* **144,** 444–450.

Bloom, P. R., and Erich, M. S. (1987). Effect of solution composition on the rate and mechanism of gibbsite dissolution in acid solutions. *Soil Sci. Soc. Am. J.* **51,** 1131–1136.

Bolt, G. A., Summer, M. E., and Kamphort, A. (1963). A study of the equilibria between three categories of potassium in an illitic soil. *Soil Sci. Soc. Am. Proc.* **27,** 294–299.

Bowman, B. T. (1979). Method of repeated additions for generating pesticide adsorption-desorption isotherm data. *Can. J. Soil Sci.* **59,** 435–437.

Bowman, B. T., and Sans, W. W. (1985). Partitioning behavior of insecticides in soil-water systems. II. Desorption hysteresis effects. *J. Environ. Qual.* **14,** 270–273.

Boyd, G. E., Adamson, A. W., and Meyers, L. S., Jr. (1947). The exchange adsorption of ions from aqueous solutions by organic zeolites. II. Kinetics. *J. Am. Chem. Soc.* **69,** 2836–2848.

Broadbent, F. E., and Clark, F. (1965). Denitrification. *In* "Soil Nitrogen" (W. V. Bartholomew and F. E. Clark, eds.), pp. 344–359. Am. Soc. Agron., Madison, Wisconsin.

Brown, T. H., and Mahler, R. L. (1987). Sorption of silica in a Northern Idaho Palouse silt loam. *Soil Sci.* **144,** 181–189.

Bunnett, J. F. (1986). Kinetics in solution. *In* "Investigations of Rates and Mechanisms of Reactions" (C. F. Bernasconi, ed.), 4th ed., pp. 171–250. Wiley, New York.

Bunzl, K. (1974a). Kinetics of ion exchange in organic matter. II. Ion exchange during continuous addition of Pb^{2+} -ions to humic acid and peat. *J. Soil Sci.* **25,** 343–356.

Bunzl, K. (1974b). Kinetics of ion exchange in soil organic matter. III. Differential ion exchange reactions of Pb^{2+} -ions in humic acid and peat. *J. Soil Sci.* **25,** 517–532.

Bunzl, K., Schmidt, W., and Sansoni, B. (1976). Kinetics of ion exchange in soil organic matter. IV. Adsorption and desorption of Pb^{2+}, Cu^{2+}, Cd^{2+}, Zn^{2+}, and Ca^{2+} by peat. *J. Soil Sci.* **27,** 32–41.

Burns, A. F., and Barbar, S. A. (1961). Effect of temperature and moisture on exchangeable potassium. *Soil Sci. Soc. Am. Proc.* **25,** 349–352.

Busenberg, E., and Clemency, C. V. (1976). The dissolution kinetics of feldspars at 25°C and 1 atm CO_2 partial pressure. *Geochim. Cosmochim. Acta* **40,** 41–49.

Caldin, E. F. (1964). "Fast Reactions in Solution." Blackwell, Oxford.

Cameron, D. R., and Klute, A. (1977). Convective-dispersive solute transport with a combined equilibrium and kinetic adsorption model. *Water Res.* **13,** 183–188.

Carski, T. H., and Sparks, D. L. (1985). A modified miscible displacement technique for investigating adsorption–desorption kinetics in soils. *Soil Sci. Soc. Am. J.* **49,** 1114–1116.

Carski, T. H., and Sparks, D. L. (1987). Differentiation of soil nitrogen fractions using a kinetic approach. *Soil Sci. Soc. Am. J.* **51,** 314–317.

Cavallaro, N., and McBride, M. B. (1978). Copper and cadmium adsorption characteristics of selected acid and calcerous soils. *Soil Sci. Soc. Am. J.* **42,** 550–556.

Chance, B. (1974). Rapid flow methods. *In* "Investigation of Rates and Mechanisms of Reactions" (G. G. Hammes, ed.), 3rd ed., pp. 187–210. Wiley, New York.

Chang, M. L., and Thomas, G. W. (1963). A suggested mechanism for sulfate adsorption by soils. *Soil Sci. Soc. Am. Proc.* **27,** 281–283.

Chien, S. H., and Clayton, W. R. (1980). Application of Elovich equation to the kinetics of phosphate release and sorption in soils. *Soil Sci. Soc. Am. J.* **44,** 265–268.

Chien, S. H., Clayton, W. R., and McClellan, G. H. (1980). Kinetics of dissolution of phosphate rocks in soils. *Soil Sci. Soc. Am. J.* **44,** 260–264.

Cho, C. M. (1971). Convective transport of ammonium with nitrification in soils. *Can. J. Soil Sci.* **51,** 339–350.

Cho, C. M., Strong, J., and Racz, G. J. (1970). Convective transport of orthophosphate (P-3 and P-32) in several Manitoba soils. *Can. J. Soil Sci.* **50,** 303–315.

Chou, L., and Wollast, R. (1984). Study of the weathering of albite at room temperature and pressure with a fluidized bed reactor. *Geochim. Cosmochim. Acta* **48,** 2205–2217.

Chou, L., and Wollast, R. (1985). Steady-state kinetics and dissolution mechanisms of albite. *Am. J. Sci.* **285,** 963–993.

Chute, J. H., and Quirk, J. P. (1967). Diffusion of potassium from mica-like materials. *Nature (London)* **213,** 1156–1157.

Cleveland, J. M. (1970). "The Chemistry of Plutonium." Gordon & Breach, New York.

Correns, C. W. (1963). Experiments on the decomposition of silicates and discussion of chemical weathering. *Clays Clay Miner.* **10,** 443–459.

Correns, C. W., and von Engelhardt, W. (1938). Neue Untersuchungen uber die Verwitterung des Kalifeldspates. *Chem. Erde* **12,** 1–22.

Crank, J. (1976). "The Mathematics of Diffusion," 2nd ed. Oxford Univ. Press, (Clarendon), London and New York.

Crist, R. H., Oberholser, K., Shank, N., and Nguyen, M. (1981). The nature of bonding between metallic ions and algae cell walls. *Environ. Sci. Technol.* **15,** 1212–1217.

Crooks, J. E., Zetter, M. S., and Tregloan, P. A. (1970). *J. Phys. E. Sci. Instr.* **3,** 73.

Crosby, D. G. (1973). The fate of pesticides in the environment. *Annu. Rev. Plant Physiol.* **24,** 467–492.

Czerlinski, V. G. (1960). Eine Versuchsanordnung zur Anwendung der Temperaturssprunsmethode auf biologische systeme. *Z. Elektrochem.* **64,** 78–79.

Dalal, R. C. (1974). Desorption of soil phosphate by anion exchange resin. *Commun. Soil Sci. Plant Anal.* **5,** 531–538.

Davidson, J. M., and Rao, P. S. C. (1978). Use of mathematical relationships to describe the behavior of nitrogen in the crop root zone. *Proc. Nat. Conf. Manage. Nitrogen Irrig. Agric.,* pp. 291.

Davidson, J. M., Graetz, D. A., Rao, P. S. C., and Selim, H. M. (1978a). Simulation of nitrogen movement, transformation, and uptake in the plant root zone. *U.S. Environ. Prot. Agency, Off. Res. Dev. [Rep.] EPA* **EPA-600/3-78-029.**

Davidson, J. M., Rao, P. S. C., and Jessup, R. E. (1978b). Critique of computer simulation modeling for nitrogen in irrigated croplands. *In* "Nitrogen in the Environment" (D. R. Nielsen and J. G. MacDonald, eds.), Vol. 1, pp. 131–143. Academic Press, New York.

Davis, J. S., and Gutfreund, H. (1976). The scope of moderate pressure changes for kinetic and equilibrium studies of biochemical systems. *FEBS Lett* **72,** 199–207.

De Camargo, O. A., Bigger, J. W., and Nielsen, D. R. (1979). Transport of inorganic phosphorus in an Alfisol. *Soil Sci. Soc. Am. J.* **43,** 884–890.

DeMaeyer, L. C. M. (1969). Electric field methods, *In* "Methods in Enzymology" (K. Kustin, ed.), Vol. 16, pp. 80–118. Academic Press, New York.

Denbigh, K. G. (1944). Velocity and yield in continuous reaction systems. *Trans. Faraday Soc.* **40,** 352–373.

Denbigh, K. G. (1966). "The Principles of Chemical Equilibrium with Applications in Chemistry and Chemical Engineering." Cambridge Univ. Press, London and New York.

Denbigh, K. G. (1981). "The Principles of Chemical Equilibrium." Cambridge Univ. Press, London and New York.

Deuel, L. E., and Swoboda, A. R. (1972). Arsenic solubility in a reduced environment. *Soil Sci. Soc. Am. Proc.* **36,** 276–278.

Di Toro, D. M., and Horzempa, L. M. (1982). Reversible and resistant components of PCB adsorption–desorption: Isotherms. *Environ. Sci. Technol.* **16,** 594–602.

Donigan, A. S., Jr., and Crawford, N. H. (1976). Modeling pesticides and nutrients on agricultural lands. *U.S. Environ. Prot. Agency, Off. Res. Dev. [Rep.] EPA* **EPA-600/2-76-043.**

Draper, N. R., and Smith, H. (1981). "Applied Regression Analysis," 2nd Ed. Wiley, New York.

Dutt, G. R., Shaffer, M. J., and Moore, W. J. (1972). Computer simulation model of dynamic bio-physicochemical processes in soils. *Ariz. Agric. Exp. Stn., Tech. Bull.* **196.**

Eigen, M. (1954). Ionic reactions in aqueous solutions with half-times as short as 10^{-9} second. Applications to neutralization and hydrolysis reactions. *Discuss. Faraday Soc.* **17,** 194–205.

Eigen, M., and DeMaeyer, L. (1963). Relaxation methods. *Tech. Org. Chem.* **8** (2), 895–1054.

Elkhatib, E. A., Bennett, O. L., and Wright, R. J. (1984a). Kinetics of arsenite sorption in soils. *Soil Sci. Soc. Am. J.* **48,** 758–762.

Elkhatib, E. A., Bennett, O. L., and Wright, R. J. (1984b). Arsenite sorption and desorption in soils. *Soil Sci. Soc. Am. J.* **48,** 1025–1030.

Enfield, C. G. (1978). Evaluation of phosphorus models for prediction of percolate water quality in land treatment. *In* "State of the Knowledge in Land Treatment of Wastewater" (H. L. McKim, ed.), Vol. 1, pp. 153–162. U.S. Army Cold Reg. Res. Eng. Lab.) Hanover, New Hampshire.

Enfield, C. G., Harlin, C. C., Jr., and Bledsoe, B. E. (1976). Comparison of five kinetic models for orthophosphate reactions in mineral soils. *Soil Sci. Soc. Am. J.* **40,** 243–249.

Evans, C. M., and Sparks, D. L. (1983). Kinetics and mechanisms of boron sorption and desorption in Mid-Atlantic Coastal Plain soils. *Agro. Abstr.* p. 146.

Evans, R. L., and Jurinak, J. J. (1976). Kinetics of phosphate release from a desert soil. *Soil Sci.* **121,** 205–211.

Eyring, E. M., and Hemmes, P. (1986). Electric field methods. *In* "Investigation of Rates and Mechanisms of Reactions" (C. F. Bernasconi, ed.), 4th ed., pp. 219–246. Wiley, New York.

Eyring, H., Liu, S. H., and Liu, S. M. (1980). "Basic Chemical Kinetics." Wiley, New York.

Farmer, W. J., and Aochi, Y. (1974). Picloram sorption by soils. *Soil Sci. Soc. Am. Proc.* **38,** 418–423.

Fava, A., and Eyring, H. (1956). Equilibrium and kinetics of detergent adsorption—A generalized equilibration theory. *J. Phys. Chem.* **60,** 890–898.

Feigenbaum, S., and Levy, R. (1977). Potassium release in some saline soils of Israel. *Geoderma* **19** 159–169.

Feigenbaum, S., Edelstein, R., and Shainberg, I. (1981). Release rate of potassium and structural cations from micas to ion exchangers in dilute solutions. *Soil Sci. Soc. Am. J.* **45,** 501–506.

Fiskell, J. G. A., Mansell, R. S., Selim, H. M., and Martin, F. G. (1979). Kinetic behavior of phosphate sorption by acid, sandy soil. *J. Environ. Qual.* **8,** 579–584.

Fokin, A. D., and Chistova, Y. D. (1967). Possibility of using internal diffusion equations to describe the kinetics of sorption processes in soils. *Sov. Soil Sci. (Engl. Transl.)* pp. 776–780.

Frank-Kamenetskii, D. A. (1979). "Diffusion and Heat Transfer in Chemical Kinetics." Plenum, New York.

Frissel, M. J., and van Veen, J. A. (1981). Simulation model for nitrogen immobilization and mineralization. *In* "Modeling Wastewater Renovation by Land Disposal" (I. K. Iskandar, ed.), pp. 359–381. Wiley, New York.

Froment, G. F., and Bischoff, K. (1979). "Chemical Reactor Analysis and Design." Wiley, New York.

Frost, A. A., and Pearson, R. G. (1961). "Kinetics and Mechanism." Wiley, New York.

Garrels, R. M. (1959). Rates of geochemical reactions at low temperatures and pressures. *In* "Research in Geochemistry" (P. H. Abelson, ed.), pp. 25–37. Wiley, New York.

Garrels, R. M., and Howard, P. (1959). Reactions of feldspar and mica with water at low temperature and pressure. *Clays Clay Miner., Proc. Conf.* **6,** 68–88.

Gedroiz, K. K. (1914). Colloidal chemistry as related to soil science. II. Rapidity of reaction exchange in the soil, colloidal condition of the soil saturated with various bases and the indicator method of determining the colloidal content of the soil. *Zh. Opytn. Agron.* **15,** 181–208.

Gettins, W. J., and Wyn-Jones, E., eds. (1979). "Techniques and Applications of Fast Reactions in Solution." Reidel Publ., Dordrecht, The Netherlands.

Gibson, Q. H. (1969). Rapid mixing: Stopped flow. *In* "Methods in Enzymology" (K. Kustin, ed.), Vol. 16, pp. 187–228. Academic Press, New York.

Glaski, F. A., and Dranoff, J. S. (1963). Ion exchange kinetics: A comparison of models. *AIChE J.* **9,** 426–431.

Glasstone, S., Laidler, K. J., and Eyring, H. (1941). "The Theory of Rate Processes." McGraw-Hill, New York.

Goldsack, D. E., Hurst, R .E., and Love, J. (1969). A pressure jump apparatus with optical detection. *Anal. Biochem.* **28,** 273–281.

Gonzalez, J. L., Herraez, M. A., and Rodriguez, S. (1982). Kinetic treatment of adsorption–desorption processes of Fe(III)–silica gel system by a theoretical generalized model. *J. Colloid Interface Sci.* **88,** 313–318.

Goring, C. A. I., and Hamaker, J. W., eds. (1972). "Organic Chemicals in the Soil Environment." Dekker, New York.

Griffin, R. A., and Burau, R. G. (1974). Kinetic and equilibrium studies of boron desorption from soil. *Soil Sci. Soc. Am. Proc.* **38,** 892–897.

Griffin, R. A., and Jurinak, J. J. (1974). Kinetics of the phosphate interaction with calcite. *Soil Sci. Soc. Am. Proc.* **38,** 75–79.

Gruenewald, B., and Knoche, W. (1979). Recent developments and applications of pressure jump methods. *In* "Techniques and Applications of Fast Reactions in Solutions" (W. J. Gettins and E. Wyn-Jones, eds.), pp. 87–94. Reidel Publ., Dordrecht, The Netherlands.

Guenzi, W. D., ed. (1974). "Pesticides in Soil and Water." Soil Sci. Soc. Am., Madison, Wisconsin.

Hachiya, K., Ashida, M. Sasaki, M., Kan, H., Inoue, T., and Yasunaga, T. (1979). Study of the kinetics of adsorption–desorption of Pb^{2+} on a γ-Al_2O_3 surface by means of relaxation techniques. *J. Phys. Chem.* **83,** 1866–1871.

Hachiya, K., Ashida, M., Sasaki, M., Karasuda, M., and Yasunaga, T. (1980). Study of the adsorption–desorption of IO_3^- pm a TiO_2 surface by means of relaxation techniques. *J. Phys. Chem.* **84,** 2292–2296.

Hachiya, K., Sasaki, M., Mikami, N., and Yasunaga, T. (1984a). Static and kinetic studies of adsorption-desorption of metal ions on a γ-Al_2O_3 surface. 1. Static study of adsorption–desorption. *J. Phys. Chem.* **88,** 23–27.

Hachiya, K., Sasaki, M., Ikeda, T., Mikami, N., and Yasunaga, T. (1984b). Static and kinetic studies of adsorption–desorption of metal ions on γAl_2O_3 surface. 2. Kinetic study by means of pressure-jump technique. *J. Phys. Chem.* **88,** 27–31.

Hackerman, N., and Stephens, S. J. (1954). The adsorption of sulfate ions from aqueous solutions by iron surfaces. *J. Phys. Chem.* **58,** 904–908.

Hamaker, J. W. (1966). Mathematical predictions of cumulative levels of pesticide in soil. *Adv. Chem. Ser.* **60,** 122–131.

Hamaker, J. W. (1972). Decomposition: Quantitative aspects. *In* "Organic Chemicals in the Environment" (C. A.I. Goring and J. W. Hamaker, eds.) pp. 255–340. Dekker, New York.

Hamaker, J. W., and Thompson, J. M. (1972). Adsorption. *In* Organic Chemicals in the Environment" (C. A.I. Goring and J. W. Hamaker, eds.) pp. 39–151. Dekker, New York.

Hamaker, J. W., Goring, C. A. I., and Youngson, C. R. (1966). Sorption and leaching of 4-amino-3,5,6-trichloropicolinic acid in soils. Organic pesticides in the environment. *Adv. Chem. Ser.* **60,** 23–37.

Hammes, G. G. (1978). "Principles of Chemical Kinetics." Academic Press, New York.

Hammett, L. P. (1960). "Physical Organic Chemistry" 2nd Edition. McGraw-Hill, New York.

Hance, R. J. (1967). Speed of attainment of sorption equilibria in some systems involving herbicides. *Weed Res.* **7,** 29–36.

Hance, R. J., ed. (1981). "Interactions Between Herbicides and the Soil." Academic Press, New York.

Haque, R., Lindstrom, F. T., Freed, V. H., and Sexton, R. (1968). Kinetic study of the sorption of 2,4-D on some clays. *Environ. Sci. Technol.* **2,** 207–211.

Haque, R., and Sexton, R. (1968). Kinetic and equilibrium study of the adsorption of 2,4-Dichlorophenoxy acetic acid on some surfaces. *J. Colloid Interface Sci.* **27,** 818–827.

Harter, R. D., and Lehmann, R. G. (1983). Use of kinetics for the study of exchange reactions in soils. *Soil Sci. Soc. Am. J.* **47,** 666–669.

Harter, R. D., and Smith, G. (1981). Langmuir equation and alternate methods of studying adsorption reactions in soils. *In* "Chemistry in Soil Environment" (R. H. Dowdy, ed.), pp. 167–182. *Am. Soc. Agronomy Special Publ.* **40,** Am. Soc. Agronomy, Madison, Wisconsin.

Havlin, J. L., and Westfall, D. G. (1985). Potassium release kinetics and plant response in calcareous soils. *Soil Sci. Soc. Am. J.* **49,** 366–370.

Havlin, J. L., Westfall, D. G., and Olsen, S. R. (1985). Mathematical models for potassium release kinetics in calcareous soils. *Soil Sci. Soc. Am. J.* **49,** 371–376.

Hayes, K. F., and Leckie, J. O. (1986). Mechanism of lead ion adsorption at the goethite–water interface. *ACS Symp. Ser.* **323,** 114–141.

Helfferich, F. (1956). Kinetik des ionenanstauschs. *Angew. Chem.* **68,** 693–698.

Helfferich, F. (1962a). "Ion Exchange." No. 2003414. University Microfilms International, Ann Arbor, Michigan.

Helfferich, F. (1962b). Ion exchange kinetics. III. Experimental test of the theory of particle-diffusion controlled ion exchange. *J. Phys. Chem.* **66,** 39–44.

Helfferich, F. (1963). Ion exchange kinetics. IV. Demonstration of the dependence of the interdiffusion coefficient on ionic composition. *J. Phys. Chem.* **67,** 1157–1158.

Helfferich, F. (1965). Ion exchange kinetics. V. Ion exchange accompanied by reactions. *J. Phys. Chem.* **69,** 1178–1187.

Helfferich, F. (1966). Ion exchange kinetics. "Ion Exchange" (J. A. Marinsky, ed.), Vol. 1, pp. 65–100. Dekker, New York.

Helfferich, F. (1983). Ion exchange kinetics–Evolution of a theory. *In* "Mass Transfer and Kinetics of Ion Exchange" (L. Liberti and F. Helfferich, eds.), pp. 157–179. Martinus Nijhoff Publ., Dordrecht, The Netherlands.

Helfferich, F., and Plesset, M. S. (1958). Ion exchange kinetics. A non-linear diffusion problem. *J. Chem. Phys.* **28,** 418–424.

Helgeson, H. C., Murphy, W. M., and Aagaard, P. (1984). Thermodynamic and kinetic constraints on reaction rates among minerals and aqueous solution. II. Rate constants, effective surface area, and the hydrolysis of feldspar. *Geochim. Cosmochim. Acta* **48,** 2405–2432.

Helling, C. S., Kearny, P. C., and Alexander, M. (1971). Behavior of pesticides in soils. *Adv. Agron.* **23,** 147–240.

Hemmes, P. (1979). Electric field methods. *In* "Techniques and Applications of Fast Reactions in Solutions" (W. J. Gettins and E. Wyn-Jones, eds.), Vol. 50, pp. 95–101. Reidel Publ. Dordrecht, The Netherlands.

Hingston, F. J., and Raupach, M. (1967). The reaction between monosilicic acid and aluminum hydroxide. 1. Kinetics of adsorption of silicic acid by aluminum hydroxide. *Aust. J. Soil Res.* **5,** 295–309.

Hissink, D. J. (1924). Base exchange in soils. *Trans. Faraday Soc.* **20,** 551–566.

Hodges, S. C., and Johnson, G. (1987). Kinetics of sulfate adsorption and desorption by Cecil soil using miscible displacement. *Soil Sci. Soc. Am. J.* **51,** 323–331.

Hoffman, H., and Pauli, K. (1966). Eine Druckwellenapparatur zur Messung von Relaxationsvorgangen in Losungen. *Ber. Busenges. Phys. Chem.* **70,** 1052–1056.

Hoffman, H., Stuehr, J., and Yeager, E. (1966). Studies of relaxation effects in electrolytic solutions with the pressure-step methods. *In* "Chemical Physics of Ionic Solutions" (B. E. Conway and R. G. Barradas, eds.), pp. 255–227. Wiley, New York.

Hoinkins, E., Levi, H. W., Lutze, W., Milkeley, N., and Tamberg, T. (1967). Isotopenaustausch-Untersuchungen an Silicat-Ionenaustauschern. IV. Zur Methodik der Bestimmung von Selbstdiffusions-koeffizienten durch heterogenen Isotopenaustausch.

Holdren, G. R., Jr., and Berner, R. A. (1979). Mechanism of feldspar weathering. I. Experimental studies. *Geochim. Cosmochim. Acta* **43,** 1161–1171.

Holdren, G. R., and Speyer, P. M. (1985). pH-dependent changes in the rates and stoichiometry of dissolution of an alkali feldspar at room temperature. *Am. J. Sci.* **285,** 994–1026.

Holdren, G. R., Jr., and Speyer, P. M. (1986). Stoichiometry of alkali feldspar dissolution at room temperature and various pH values. *In* "Rates of Chemical Weathering of Rocks and Minerals" (S. M. Colman and D. P. Dethier, eds.), pp. 61–81. Academic Press, Orlando, Florida.

Holdren, G. R., and Speyer, P. M. (1987). Reaction rate–surface area relationships during the early stages of weathering. II. Data on eight additional feldspars. *Geochim, Cosmochim. Acta* **51,** 2311–2318.

Horzempa, L. M., and DiToro, D. M. (1983). The extent of reversibility of polychlorinated biphenyl adsorption. *Water Res.* **17,** 851–859.

Huang, P. M., and Liaw, W. K. (1979). Adsorption of arsenite by lake sediments. *Int. Rev. Gesamten Hydrobiol.* **64,** 263–271.

Huang, P. M., Crossan, L. S., and Rennie, D. A. (1968). Chemical dynamics of K release from potassium minerals common in soils. *Trans., Int. Congr. Soil Sci., 9th, 1968* Vol. 2, pp. 705–712.

Huang, W. H., and Kiang, W. C. (1972). Laboratory dissolution of plagioclase feldspars in water and organic acids at room temperature. *Am. Mineral.* **57,** 1849–1859.

Ikeda, T., and Yasunaga, T. (1984). Kinetics studies of ion exchange of NH_4^+ for H^+ in zeolite H-ZSM-5 by the chemical relaxation method. *J. Colloid Interface Sci.* **99,** 183–186.

Ikeda, T., Sasaki, M., Astumian, R. D., and Yasunaga, T. (1981). Kinetics of the hydrolysis of zeolite 4A surface by the pressure-jump relaxation method. *Bull. Chem. Soc. Jpn.* **54,** 1885–1886.

Ikeda, T., Sasaki, M., and Yasunaga, T. (1982a). Kinetics of the hydrolysis of hydroxyl groups on zeolite surfaces using the pressure-jump relaxation method. *J. Phys. Chem.* **86,** 1678–1680.

Ikeda, T., Sasaki, M., Hachiya, K., Astumian, R. D., Yasunaga, T., and Schelly, Z. A. (1982b). Adsorption–desorption kinetics of acetic acid on silica–aluminum particles in aqueous suspensions, using p-jump relaxation method. *J. Phys. Chem.* **86,** 3861–3866.

Ikeda, T., Sasaki, M., and Yasunaga, T. (1983). Kinetic studies of ion exchange of alkylammonium ion for sodium ion in aqueous suspensions of zeolite 4A using the pressure-jump method. *J. Phys. Chem.* **87,** 745–749.

Ikeda, T., Nakahara, J., Sasaki, M., and Yasunaga, T. (1984a). Kinetic behavior of alkali metal ion on zeolite 4A surface using the stopped-flow method. *J. Colloid Interface Sci.* **97,** 278–283.

Ikeda, T., Sasaki, M., and Yasunaga, T. (1984b). Kinetic studies of ion exchange of NH_4^+ in zeolite H-ZSM-5 by the chemical relaxation method. *J. Colloid Interface Sci.* **98,** 192–195.

Ilgenfritz, G. (1966). Ph.D. Dissertation, George August University, Goettingen.

Iskandar, I. K. ed. (1981). "Modeling Wastewater Renovation." Wiley, New York.

Jardine, P. M., and Sparks, D. L. (1984). Potassium–calcium exchange in a multireactive soil system. I. Kinetics. *Soil Sci. Soc. Am. J.* **48,** 39–45.

Jardine, P. M., and Zelazny, L. W. (1986). Mononuclear and polynuclear aluminum speciation through differential kinetic reactions with ferron. *Soil Sci. Soc. Am. J.* **50,** 895–900.

Jardine, P. M., Zelazny, L. W., and Parker, J. C. (1985a). Mechanisms of aluminum adsorption on clay minerals and peat. *Soil Sci. Soc. Am. J.* **49,** 862–867.

Jardine, P. M., Parker, J. C., and Zelazny, L. W. (1985b). Kinetics and mechanisms of aluminum adsorption on kaolinite using a two-site nonequilibrium transport model. *Soil Sci. Soc. Am. J.* **49,** 867–873.

Johnson, D. L. (1972). Bacterial reduction of arsenate in sea water. *Nature (London)* **240,** 44–45.

Johnson, D. L., and Pilson, M. E. A. (1975). The oxidation of arsenite in seawater. *Environ. Lett.* **8,** 157–171.

Jopony, M., and Young, S. D. (1987). A constant potential titration method for studying the kinetics of Cu^{2+} desorption from soil and clay minerals. *J. Soil Sci.* **38,** 219–228.

Jost, von A. (1966). SoBwellen in Flussigkeiten zur Untersuchung von Losungsreakitionen. *Ber. Bunsenges. Phys. Chem.* **70,** 1057–1060.

Karickhoff, S. W. (1980). Sorption kinetics of hydrophobic pollutants in natural sediments. *In* "Contaminants and Sediments: Analysis, Chemistry, Biology," (R. A. Baker, ed.), Vol. 2, pp. 193–205. Ann Arbor Sci. Publ., Ann Arbor, Michigan.

Karickhoff, S. W., Brown, D. S., and Scott, T. A. (1979). Sorption of hydrophobic pollutants on natural sediments. *Water Res.* **13,** 241–248.

Kaufman, D. D. (1976). Soil degradation and persistence of benchmark pesticides. *In* "A Literature Survey of Benchmark Pesticides," pp. 19–71. George Washington Univ. Medical Center, Washington, D.C.

Kearny, P. C., and Helling, C. S. (1969). Reactions of pesticides in soils. *Residue Rev.* **25,** 25–44.

Keay, J., and Wild, A. (1961). The kinetics of cation exchange in vermiculite. *Soil Sci.* **92,** 49–54.

Keeney, D. R. (1973). The nitrogen cycle in sediment water systems. *J. Environ. Qual.* **2,** 15–29.

Kelley, W. P. (1948). Cation exchange in soils. *ACS Monogr.* **109.**

Kennedy, V. C., and Brown, T. C. (1965). Experiments with a sodium-ion electrode as a means of studying cation exchange rates. *Clays Clay Miner.* **13,** 351–352.

Khan, S. U. (1973). Equilibrium and kinetic studies on the adsorption of 2,4-D and picloram on humic acid, *Can. J. Soil Sci.* **53,** 429–434.

Kinshoffer, G. S., Bunzl, K., Sansoni, B., and Schwab, G. M. (1973). Differential ion exchange in a zeolite. *Z. Phys. Chem.* **87,** 218–232.

Knoche, W. (1974). Pressure-jump methods. *In* "Investigations of Rates and Mechanisms of Reactions" (G. G. Hammes, ed.), 3rd ed., pp. 187–210. Wiley, New York.

Knoche, W., and von Strehlow, H. (1979). Data capture and processing in chemical relaxation measurements. *In* "Techniques and Applications of Fast Reactions in Solution" (W. J. Gettins and E. Wyn-Jones, eds.), pp. 137–142. Reidel Publ., Dordrecht, The Netherlands.

Knoche, W., and Wiese, G. (1974). An improved apparatus for pressure-jump relaxation measurements. *Chem. Instrum. (N.Y.)* **5,** 91–98.

Knoche, W., and Wiese, G. (1976). Pressure-jump relaxation technique. *Rev. Sci. Instrum.* **47,** 220–221.

Kohl, D. H., Vithayathil, F., Whitlow, R., Shearer, G., and Chien, S. H. (1976). Denitrification kinetics in soil systems: The significance of good fits of data to mathematical forms. *Soil Sci. Soc. Am. J.* **40,** 249–253.

Komareni, S. (1978). Cesium sorption and desorption behavior of kaolinites. *Soil Sci. Soc. Am. J.* **42,** 531–532.

Kressman, T. R. E., and Kitchener, J. A. (1949). Cation exchange with a synthetic phenolsulphonate resin, V. Kinetics. *Discuss. Faraday Soc.* **7,** 90–103.

Krizan, M., and von Strehlow, H. (1974). On the evaluation of chemical relaxation measurements with sampling techniques and on-line processing. *Chem. Instrum. (N.Y.)* **5,** 99–108.

Kuo, S., and Lotse, E. G. (1972). Kinetics of phosphate adsorption by calcium carbonate and Ca-kaolinite. *Soil Sci. Soc. Am. Proc.* **36,** 725–729.

Kuo, S., and Lotse, E. G. (1973). Kinetics of phosphate adsorption and desorption by hematite and gibbsite. *Soil Sci.* **116,** 400–406.

Kuo, S., and Lotse, E. G. (1974). Kinetics of phosphate adsorption and desorption by lake sediments. *Soil Sci. Soc. Am. Proc.* **38,** 50–54.

Kustin, K., ed. (1969). "Methods in Enzymology," Vol. 16, Academic Press, New York.

Lagache, M. (1965). Contribution á l'étude de l'altération des feldspaths, dans l'eau, entre 100 et 200° sous diverses pressions de CO_2, et application á la synthése des minéraux argileux. *Bull. Soc. Fr. Mineral. Cristallogr.* **88,** 223–253.

Laidler, K. J. (1965). "Chemical Kinetics." McGraw-Hill, New York.

Laveglia, J., and Dahm, P. A. (1977). Degradation of organophosphorus and carbamate insecticides in the soil and by microorganisms. *Annu. Rev. Entomol.* **22,** 483–513.

Leenheer, J. A., and Ahlrichs, J. L. (1971). A kinetic and equilibrium study of the adsorption of carbaryl and parathion upon soil organic matter surfaces. *Soil Sci. Soc. Am. Proc.* **35,** 700–704.

Levenspiel, O. (1972). "Chemical Reaction Engineering." Wiley, New York.

Li, W. C., Armstrong, D. E., Williams, J. D. H., Harris, R. F., and Syers, J. K. (1972). Rate and extent of inorganic phosphate exchange in lake sediments. *Soil Sci. Soc. Am. Proc.* **36,** 279–285.

Liberti, L., and Passino, R. (1983). Physical and chemical kinetics in ion exchange. *Gazz. Chim. Ital.* **113,** 627–633.

Liberti, L., and Passino, R. (1985). Ion exchange kinetics in selective systems. *In* "Ion Exchange and Solvent Extraction" (J. A. Marinsky and Y. Marcus, eds.), pp. 175–210. Dekker, New York.

Lin, C., Busscher, W. J., and Douglas, L. A. (1983). Multifactor kinetics of phosphate reactions with minerals in acid soils. II. Experimental curve fitting. *Soil Sci. Soc. Am. J.* **47,** 1103–1109.

Liu, M., and Thomas, G. W. (1961). Nature of sulfate retention by acid soils. *Nature (London)* **192,** 384.

Lindstrom, F. T., Haque, R., and Coshow, W. R. (1970). Adsorption from solution. II. A new model for the kinetics of adsorption–desorption processes. *J. Phys. Chem.* **74,** 495–502.

Ljunggren, S., and Lamm. (1958). A relaxation method for the determination of moderately rapid reaction rates near chemical equilibrium. *Acta Chem. Scand.* **12,** 1834–1850.

Low, M. J. D. (1960). Kinetics of chemisorption of gases on solids. *Chem. Rev.* **60,** 267–312.

Luce, R. W., Bartlett, R. W., and Parks, G. A. (1972). Dissolution kinetics of magnesium silicates. *Geochim. Cosmochim Acta.* **36,** 35–40.

Lutze, W., and Mickeley, N. (1971). Ion exchange kinetics in vermiculite. *J. Phys. Chem.* **75,** 2484–2488.

McCall, P. J., and Agin, G. L. (1985). Desorption kinetics of picloram as affected by residence time in the soil. *Environ. Toxicol. Chem.* **4,** 37–44.

Macri, G., and Petrucci, S. (1970). Pressure jump relaxation kinetics of magnesium(II), manganese(II), nickel(II), cobalt(II), copper(II), and zinc(II) *m*-benzenedisulfonates in anhydrous methanol at 25°. *Inorg. Chem.* **9,** 1009–1014.

Malcom, R. L., and Kennedy, V. C. (1969). Rate of cation exchange on clay minerals as determined by specific ion electrode techniques. *Soil Sci. Am. Proc.* **33,** 247–253.

Malcom, R. L., and Kennedy, V. C. (1970). Variation of cation exchange capacity and rate with particle size in stream sediment. *J. Water Pollut. Control Fed.* **42,** 153–160.

Mansell, R. S., Selim, H. M., and Fiskell, J. G. A. (1977a). Simulated transformations and transport of phosphorus in soil. *Soil Sci.* **124,** 102–109.

Mansell, R. S., Selim, H. M., Kanshanasut, P., Davidson, J. M., and Fiskell, J. G. A. (1977b). Experimental and simulated transport of phosphorus through sandy soils. *Water Res.* **13,** 189–194.

Mansell, R. S., and Selim, H. M. (1981). Mathematical models for predicting reactions and transport of phosphorus applied to soils. *In* "Modeling Wastewater Renovation" (R. S. Iskandar, ed.), pp. 600–646. Wiley, New York.

Mehran, M., and Tanji, K. K. (1974). Computer modeling of nitrogen transformations in soils. *J. Environ. Qual.* **3,** 391–395.

Mikami, N., Sasaki, M., Hachlya, K., Astumian, R. D., Ikeda, T., and Yasunaga, T. (1983a). Kinetics of the adsorption of PO_4 on the γ-Al_2O_3 surface using the pressure-jump technique. *J. Phys. Chem.* **87,** 1454–1458.

Mikami, N., Sasaki, M., Kikuchi, T., and Yasunaga, T. (1983b). Kinetics of adsorption–desorption of chromate on γ-Al_2O_3 surfaces using the pressure-jump technique. *J. Phys. Chem.* **87,** 5245–5248.

Miller, D. M., Miller, W. P., and Sumner, M. E. (1986). Kinetics of silicic acid sorption by goethite using a flow-through cell. *Agron. Abstr.* p. 169.

Molz, F. J., and Remson, I. (1970). Extraction-term models of soil moisture use by transpiring plants. *Water Res.* **6,** 1346–1356.

Montgomery, D. C., and Peck, E. A. (1982). "Introduction to Linear Regression Analysis." Wiley, New York.

Moore, J. W., and Pearson, R. G. (1981). "Kinetics and Mechanism." Wiley, New York.

Morgan, J. J., and Stumm, W. (1964). Colloid-chemical properties of manganese dioxide. *J. Colloid Sci.* **19,** 347–359.

Mortland, M. M. (1958). Kinetics of potassium release from biotite. *Soil Sci. Soc. Am. Proc.* **22,** 503–508.

Mortland, M. M. (1970). Clay-organic complexes and interactions, *Adv. Agron.* **22,** 75–117.

Mortland, M. M., and Ellis, B. G. (1959). Release of fixed potassium as a diffusion-controlled process. *Soil Sci. Soc. Am. Proc.* **23,** 363–364.

Mukhtar, M. (1976). Desorption of adsorbed ametryn and diuron from soils and soil components in relation to rates, mechanisms, and energy of adsorption reactions. Ph.D. Dissertation, University of Hawaii, Honolulu.

Murray, D. S., Santelman, P. W., and Davidson, J. M. (1975). Comparative adsorption, desorption, and mobility of dipropetryn and prometryn in soil. *J. Agric. Food Chem.* **23,** 578–582.

Negishi, H., Sasaki, M., Iwaki, T., Hayes, K. F., and Yasunaga, T. (1984). Kinetic study of adsorption–desorption of methanol on H-ZSM-5 using a new "gas-concentration jump" technique. *J. Phys. Chem.* **88,** 5564–5569.

Nelson, P. O., Chung, A. K., and Hudson, M. C. (1981). Factors affecting the fate of heavy metals in the activated sludge process. *J. Water Pollut. Control Fed.* **53,** 1323–1333.

Nernst, W. (1904). Theorie der Reaktionsgeschwindigkeit in heterogenen systemen. *Z. Phys. Chem.* **47,** 52–55.

Neufeld, R. D., and Heiman, E. R. (1975). Heavy metal removal by acclimated activated sludge. *J. Water Pollut. Control Fed.* **47,** 310–329.

Nkedi-Kizza, P., Biggar, J. W., Selim, H. M., VanGenuchten, M. N., Wierenga, P. J., Davidson, J. M., and Nielsen, D. R. (1984). On the equivalence of two conceptual models for describing ion exchange during transport through an aggregated Oxisol, *Water Resour. Res.* **20,** 1123–1130.

Noll, M. R., Bertsch, P. M., and Sparks, D. L. (1986). Kinetics of Cs^+–K^+ and Cs^+–NH_4^+ exchange on kaolinite and vermiculite. *Agron. Abstr.* p. 170.

Novak, L. T., and Adriano, D. C. (1975). Phosphorus movement in soils: Soil–orthophosphate reaction kinetics. J. Environ. Qual. **4,** 261–266.

Ogwada, R. A., and Sparks, D. L. (1986a). A critical evaluation on the use of kinetics for determining thermodynamics of ion exchange in soils. *Soil Sci. Soc. Am. J.* **50,** 300–305.

Ogwada, R. A., and Sparks, D. L. (1986b). Kinetics of ion exchange on clay minerals and soil. I. Evaluation of methods. *Soil Sci. Soc. Am. J.* **50,** 1158–1162.

Ogwada, R. A., and Sparks, D. L. (1986c). Kinetics of ion exchange on clay minerals and soil. II. Elucidation of rate-limiting steps. *Soil Sci. Soc. Am. J.* **50,** 1162–1164.

Olsen, S. R., and Khasawneh, F. E. (1980). Use and limitations of physical-chemical criteria for assessing the status of phosphorus in soils. *In* "The Role of Phosphorus in Agriculture" (F. E. Khasawneh, E. C. Sample, and E. J. Kamprath, eds.), pp. 361–410. Am. Soc. Agron., Madison, Wisconsin.

Onken, A. B., and Matheson, R .L. (1982). Dissolution rate of EDTA-extractable phosphate from soils. *Soil Sci. Soc. Am. J.* **46,** 276–279.

Oscarson, D. W., Huang, P. M., and Liaw, W. K. (1980). The oxidation of arsenite by aquatic sediments. *J. Environ. Qual.* **9,** 700–703.

Oscarson, D. W., Huang, P. M., Defosse, C., and Herbillon, A. (1981a). Oxidative power of Mn(IV) and Fe(III) oxides with respect to As(III) in terrestrial and aquatic environments. *Nature (London)* **291,** 50–51.

Oscarson, D. W., Huang, P. M., and Liaw, W. K. (1981b). Role of manganese in the oxidation of arsenite by freshwater lake sediments. *Clays Clay Miner.* **29,** 219–225.

Oscarson, D. W., Huang, P. M., and Hammer, U. T. (1983a). Oxidation and sorption of arsenite by manganese dioxide as influenced by surface coatings of iron and aluminum oxides and calcium carbonate. *Water, Air, Soil, Pollut.* **20,** 233–244.

Oscarson, D. W., Huang, P. M., Liaw, W. K., and Hammer, U. T. (1983b). Kinetics of oxidation of arsenite by various manganese dioxides. *Soil Sci. Soc. Am. J.* **47,** 644–648.

Overman, A. R., and Chu, R. L. (1977a). A kinetic model of steady state phosphorus fixation in a batch reactor. I. Effect of soil/solution ratio. *Water Res.* **11,** 771–775.

Overman, A. R., and Chu, R. L. (1977b). A kinetic model of steady state phosphorus fixation in a batch reactor. II. Effect of pH. *Water Res.* **11,** 777–778.

Overman, A. R., and Chu, R. L. (1977c). A kinetic model of steady-state phosphorus fixation in a batch reactor. III. Effect of solution reaction. *Water Res.* **11,** 779–781.

Overman, A. R., Chu, R. L., and Leseman, W. G. (1976). Phosphorus transport in a packed bed reactor. *J. Water Pollut. Control Fed.* **43,** 880–888.

Overman, A. R., McMahon, B. R., and Chu, R. L. (1980). Velocity dependence of phosphorus transport in a packed-bed reactor. *J. Water Pollut. Control Fed.* **52,** 2471–2476.

Paces, T. (1973). Steady-state kinetics and equilibrium between groundwater and granitic rock. *Geochim. Cosmochim. Acta* **37,** 2641–2663.

Panda, D., Mohapatra, J. S., Misra, C., and Panda, N. (1978). Movement and retention of phosphate in Lateritic soil cores. *J. Indian Soc. Soil Sci.* **26,** 25–32.

Parravano, G., and Boudart, M. (1955). Chemisorption and catalysis on oxide semiconductors. *Adv. Catal.* **7,** 47–62.

Pasricha, N. S., Sparks, D. L., and Toner, C. V., IV (1987). Competitive anion adsorption kinetics on soils of variable charge. *Agron. Abstr.* p. 172.

Patel, R. C., Atkinson, G., and Boe, R. J. (1974). Pressure jump apparatus for the study of fast reactions. *Chem. Instrum. (N.Y.)* **5,** 243–255.

Patrick, W. H., Jr. (1961). Nitrate reduction rate in a submerged soil as affected by redox potential. *Trans. Int. Congr. Soil Sci., 7th, 1960* Vol. 2, pp. 494–500.

Peck, D. E. (1977). The adsorption–desorption of diuron by fresh water sediments. M.S. Thesis, University of California, Riverside.

Peck, D. E., Corwin, D. L., and Farmer, W. J. (1980). Adsorption–desorption of diuron by freshwater sediments. *J. Environ. Qual.* **9,** 101–106.

Peryea, F. J., Bingham, F. T., and Rhoades, J. D. (1985). Kinetics of post-reclamation boron dissolution. *Soil Sci. Soc. Am. J.* **49,** 836–839.

Petrovic, R., Berner, R. A., and Goldhaber, M. B. (1976). Rate control in dissolution of alkali feldspars. I. Study of residual feldspar grains by x-ray photoelectron spectroscopy. *Geochim. Cosmochim. Acta* **40,** 537–548.

Phelan, P. J., and Mattigod, S. V. (1987). Kinetics of heterogeneously initiated precipitation of calcium phosphates. *Soil Sci. Soc. Am. J.* **51,** 336–341.

Plesset, M. S., Helfferich, F., and Franklin, J. N. (1958). Ion exchange kinetics. A nonlinear diffusion problem. II. Particle diffusion controlled exchange of univalent and bivalent ions. *J. Chem. Phys.* **29,** 1064–1069.

Plicka, J., Cabicar, J., Stamberg, K., and Favian, M. (1984). The kinetics of ion exchange sorption in ternary systems. *In* "Ion Exchange Technology" (D. Naden and M. Streat, eds.), pp. 331–336. Dekker, New York.

Polyzopoulos, N. A., Keramidas, V. Z., and Pavlatou, A. (1986). On the limitations of the simplified Elovich equation in describing the kinetics of phosphate sorption and release from soils. *J. Soil Sci.* **37,** 81–87.

Rai, E., and Serne, R. J. (1977) Plutonium activities in soil solutions and the stability and formation of selected plutonium minerals. *J. Environ. Qual.* **6,** 89–95.

Rajan, S. S. S. (1978). Sulfate adsorbed on hydrous alumina, ligands displaced, changes in surface charge. *Soil Sci. Soc. Am. J.* **42,** 39–44.

Randle, K., and Hartmann, E. H. (1987). Applications of the continuous flow stirred cell (CFSC) technique to adsorption of zinc, cadmium and mercury on humic acids, *Geoderma* **40,** 281–296.

Rao, P. S. C. (1974). Pore-geometry effects on solute dispersion in aggregated soils and evaluation of a predictive model. Ph.D. Dissertation, University of Hawaii, Honolulu.

Rao, P. S. C., and Davison, J. M. (1980). Estimation of pesticide retention and transformation parameters required in nonpoint source pollution models. *In* "Environmental Impact of Nonpoint Source Pollution" (M. R. Overcash and J. M. Davidson, eds.), pp. 23–67. Ann Arbor Sci. Publ., Ann Arbor, Michigan.

Rao, P. S. C., and Davison, J. M., eds (1982). "Retention and Transformation of Selected Pesticides and Phosphorus in Soil-Water Systems: A Critical Review," EPA-600/3-82-060. U.S. Environ. Prot. Agency, Athens, Georgia.

Rao, P. S. C., and Jessup, R. E. (1982). Development and verification of simulation models for describing pesticide dynamics in soils. *Ecol. Modell.* **16,** 67–75.

Rao, P. S. C., and Jessup, R. E. (1983). Sorption and movement of pesticides and other toxic organic substances in soils. *SSSA Spec. Publ.* **11,** 183–201.

Rao, P. S. C., Selim, H. M., Davidson, J. M., and Graetz, D. A. (1976). Simulation of transformation, ion exchange and transport of selected nitrogen species in soils. *Proc.—Soil Crop Sci. Soc. Fla.* **35,** 161–164.

Rao, P. S. C., Davidson, J. M., Jessup, R. E., and Selim, H. M. (1979). Evaluation of conceptual models for describing nonequilibrium adsorption–desorption of pesticides during steady-flow in soils. *Soil Sci. Soc. Am. J.* **43,** 22–28.

Reddy, K. R., Khaleel, R., Overcash, M. R., and Westerman, P. W. (1979a). A nonpoint source model for land areas receiving animal wastes. I. Mineralization of organic nitrogen. *Trans. ASAE* **22,** 863–872.

Reddy, K. R., Khaleel, R., Overcash, M. R., and Westerman, P. W. (1979b). A nonpoint source model for land areas receiving animal wastes. II. Ammonia volatilization. *Trans. ASAE* **22,** 1398–1405.

Reed, M. G., and Scott, A. D. (1962). Kinetics of potassium release from biotite and muscovite in sodium tetraphenylboron solution. *Soil Sci. Soc. Am. Proc.* **26,** 437–440.

Reichenberg. D. (1957). Properties and behavior: Kinetics, electrolyte penetration, and absorption of nonelectrolytes. *In* "Ion Exchangers in Organic and Biochemistry" (C. Calmon and L. R. E. Kressman, eds.), pp. 66–85. Wiley, New York.

Robinson, B. H. (1975). The stopped-flow and temperature-jump techniques—principles and recent advances. *In* "Chemical and Biological Applications of Relaxation Spectrometry" (E. Wyn-Jones, ed.), pp. 41–48. Reidel Publ., Dordrecht, The Netherlands.

Robinson, B. H. (1986). Rapid flow methods. *In* "Investigation of Elementary Reaction Steps in Solution and Fast Reaction Techniques" (C. F. Bernasconi, ed.), 4th ed., pp. 9–29. Wiley, New York.

Sadusky, M. C., Sparks, D. L., Noll, M. R., and Hendricks, G. J. (1987). Kinetics and mechanisms of potassium release from sandy Middle Atlantic Coastal Plain Soils. *Soil Sci. Soc. Am. J.* **51,** 1460–1465.

Salim, R., and Cooksey, B. G. (1980). Kinetics of the adsorption of lead on river mud. *Plant Soil* **54,** 399–417.

Saltzman, S., and Yaron, B., eds. (1986). "Pesticides in Soils." Van Nostrand-Reinhold, New York.

Sasaki, M., Morlya, M. Yasunaga, T., and Astumian, R. D. (1983). A kinetic study of ion-pair formation on the surface of α-FeOOH in aqueous suspensions using the electric field pulse technique. *J. Phys. Chem.* **87,** 1449–1453.

Sasaki, M., Negishi, H., Ohuchi, H., Inoue, M., and Yasunaga, T. (1985). Intercalation kinetics study of alkali-metal ions into TiS_2 using the pressure-jump technique. *J. Phys. Chem.* **89,** 1970–1973.

Sawhney, B. L. (1966). Kinetics of cesium sorption by clay minerals. *Soil Sci. Soc. Am. Proc.* **30,** 565–569.

Schlogl, R., and Helfferich, F. (1957). Comment on the significance of diffusion potentials in ion exchange kinetics. *J. Chem. Phys.* **26,** 5–7.

Schnabel, R. R., and Fitting, D. J. (1988). Analysis of chemical kinetics data from dilute, dispersed, well-mixed flow-through systems. *Soil Sci. Soc. Am. J.* **52,** in press.

Schott, J., Berner, R. A., and Lennart-Sjöberg, E. L. (1981). Mechanism of pyroxene and amphibole weathering. I. Experimental studies of Fe-free minerals. *Geochim. Cosmochim. Acta* **45,** 2123–2135.

Schwartz, A., Marinsky J. A., and Spiegler, K. S. (1964). Self-exchange measurements in a chelating ion-exchange resin. *J. Phys. Chem.* **68,** 918.

Scott, A. D., and Reed, M. G. (1962). Chemical extraction of potassium from soils and micaceous minerals with solution containing sodium tetraphenylboron. II. Biotite. *Soil Sci. Soc. Am. Proc.* **26,** 41–45.

Selim, H. M., Kanchanasut, P., Mansell, R. S., Zelazny, L. W., and Davidson, J. M. (1975). Phosphorus and chloride movement in a Spodosol. *Proc.—Soil Crop Sci. Soc. Fla.* **34,** 18–23.

Selim, H. M., Mansell, R. S., and Zelazny, L. W. (1976a). Modeling reactions and transport of potassium in soils. *Soil Sci.* **122,** 77–84.

Selim, H. M., Davidson, J. M., and Mansell, R. S. (1976a). Evaluation of a two-site adsorption–desorption model for describing solute transport in soils. *Proc. Summer Comput. Simul. Conf.* pp. 444–448.

Senesi, N., and Testini, C. (1982). Physicochemical investigations of interaction mechanisms between *s*-triazine herbicides and soil humic acids. *Geoderma* **28,** 129–146.

Seyfried, M. S., Sparks, D. L., Bar-Tal, A., and Feigenbaum, S. (1988). Kinetics of Ca–Mg exchange on soil using a stirred-flow reaction chamber. *Soil Sci. Soc. Am. J.* **52,** in review.

Shaffer, M. J., Ribbens, R. W., and Huntly, C. W. (1977). Production of Mineral Quality of Irrigation Return Flow. V. Detailed Return Flow Salinity and Nutrient Simulation Model, EPA-600/2-77-179E. U.S. Environ. Prot. Agency, Ada. Oklahoma.

Sharma, H. D., Jervis, R. E., and McMillen, L. W. (1970). Kinetics of ion exchange. Diffusion of trace components. *J. Phys. Chem.* **74,** 969–978.

Sharpley, A. N. (1983). Effect of soil properties on the kinetics of phosphorus desorption. *Soil Sci. Soc. Am. J.* **47,** 462–467.

Sharpley, A. N. (1987). The kinetics of potassium desorption. *Soil Sci. Soc. Am. J.* **51,** 912–917.

Sharpley, A. N., Ahuja, L. R., and Menzel, R. G. (1981a). The release of soil phosphorus to runoff in relation to the kinetics of desorption. *J. Environ. Qual.* **10,** 386–391.

Sharpley, A. N., Ahuja, L. R., Yamamoto, M., and Menzel, R. G. (1981b). The kinetics of phosphorus desorption from soil. *Soil Sci. Soc. Am. J.* **45,** 493–500.

Sivasubramaniam, S., and Talibudeen, O. (1972). Potassium-aluminum exchange in acid soils. I. Kinetics. *J. Soil Sci.* **23,** 163–173.

Skopp, J. (1986). Analysis of time dependent chemical processes in soils. *J. Environ. Qual.* **15,** 205–213.

Skopp. J., and McAllister, D. L. (1986). Chemical kinetics from a thin disc flow system: Theory. *Soil Sci. Soc. Am. J.* **50,** 617–623.

Skopp, J., and Warrick, A. W. (1974). A two-phase model for the miscible displacement of reactive solutes in soils. *Soil Sci. Soc. Am. J.* **45,** 493–496.

Smith, T. G., and Dranoff, J. S. (1964). Film diffusion-controlled kinetics in binary ion exchange. *Ind. Eng. Chem. Fundam.* **3,** 195–200.

Sparks, D. L. (1985). Kinetics of ionic reactions in clay minerals and soils. *Adv. Agron.* **38,** 231–266.

Sparks, D. L. (1986). Kinetics of reactions in pure and in mixed systems. *In* "Soil Physical Chemistry" (D. L. Sparks, ed.), pp. 83–178. CRC Press, Boca Raton, Florida.

Sparks, D. L. (1987a). Dynamics of soil potassium. *Adv. Soil Sci.* **6,** 1–63.

Sparks, D. L. (1987b). "Kinetics of Soil Chemical Processes: Past Progress and Future Needs." Soil Sci. Soc. Am., Madison, Wisconsin.

Sparks, D. L., and Huang, P. M. (1985). Physical chemistry of soil potassium. *In* "Potassium in Agriculture" (R. E. Munson, ed.), pp. 201–276. Am. Soc. Agronomy, Madison, Wisconsin.

Sparks, D. L., and Jardine, P. M. (1981). Thermodynamics of potassium exchange in soil using a kinetics approach. *Soil Sci. Soc. Am. J.* **45,** 1094–1099.

Sparks, D. L., and Jardine, P. M. (1984). Comparison of kinetic equations to describe K–Ca exchange in pure and in mixed systems. *Soil Sci.* **138,** 115–122.

Sparks, D. L., and Recheigl, J. E. (1982). Comparison of batch and miscible displacement techniques to describe potassium adsorption kinetics in Delaware soils. *Soil Sci. Soc. Am. J.* **46,** 875–877.

Sparks, D. L., Zelazny, L. W., and Martens, D. C. (1980a). Kinetics of potassium exchange in a Paleudult from the Coastal Plain of Virginia. *Soil Sci. Soc. Am. J.* **44,** 37–40.

Sparks, D. L., Zelazny, L. W., and Martens, D. C. (1980b). Kinetics of potassium desorption in soil using miscible displacement. *Soil Sci. Soc. Am. J.* **44,** 1205–1208.

Sposito, G. (1986). Thermodynamics of the soil solution. *In* "Soil Physical Chemistry" (D. L. Sparks, ed.), pp. 147–178. CRC Press, Boca Raton, Florida.

Stanford, G., Vanderpol, R. A., and Dzienia, S. (1975). Denitrification rates in relation to total and extractable soil carbon. *Soil Sci. Soc. Am. Proc.* **39,** 284–289.

Steinberg, S. M., Pignatello, J. J., and Sawhney, B. L. (1987). Persistence of 1,2-Dibromoethane in soils: Entrapment in intraparticle micropores. *Environ. Sci. Technol.* **21,** 1201–1208.

Stone, A. T. (1986). Adsorption of organic reductants and subsequent electron transfer on metal oxide surfaces, *ACS Symp. Ser.* **323,** 446–461.

Stone, A. T. (1987a). Microbial metabolates and the reductive dissolution of manganese oxides: Oxalate and pyruvate. *Geochim. Cosmochim. Acta* **51,** 919–925.

Stone, A. T. (1987b). Reductive dissolution of manganese(III/IV) oxides by substituted phenols. *Environ. Sci. Technol.* **21,** 979–988.

Stone, A. T., and Morgan, J. J. (1984a). Reduction and dissolution of manganese(III) and manganese (IV) oxides by organics. 1. Reaction with hydroquinone. *Environ. Sci. Technol.* **18,** 450–456.

Stone, A. T., and Morgan, J. J. (1984b). Reduction and dissolution of manganese(III) and manganese(IV) oxides by organics. 2. Survey of the reactivity of organics. *Environ. Sci. Technol.* **18,** 617–624.

Strehlow, von H., and Becker, M. (1959). Ein Drucksprung-Verfahren zur messung der Geschwindigkeit. *Z. Elektrochem.* **63,** 457–561.

Strehlow, von H., and Wendt, H. (1963). Fast ionic reactions in solution. IV. The formation of the vanadyl sulfate complex in aqueous solution. *Inorg. Chem.* **2,** 6–10.

Stumm, W. (1986). Coordinative interactions between soil solids and water—An aquatic chemist's point of view. *Geoderma* **38,** 19–30.

Stumm, W., and Morgan, J. J. (1981). "Aquatic Chemistry," 2nd ed. Wiley, New York.

Stumm, W., Furrer, G., and Kunz, B. (1983). The role of surface coordination in precipitation and dissolution of mineral phases. *Croat. Chem. Acta* **56,** 593–611.

Strumm, W., Furrer, G., Wieland, E., and Zinder, B. (1985). The effects of complex-forming ligands on the dissolution of oxides and aluminosilicates. *In* "The Chemistry of Weathering" (J. I. Drever, ed.), pp. 55–74. Reidel Publ., Dordrecht, The Netherlands.

Takahashi, M. T., and Alberty, R. A. (1969). The pressure-jump methods. *In* "Methods in Enzymology" (K. Kustin, ed.), Vol. 16, pp. 31–55. Academic Press, New York.

Talibudeen, O., and Dey, S. K. (1968). Potassium reserves in British soils. *J. Agric. Sci.* **71,** 95–104.

Tanji, K. K. (1982). Modeling of the nitrogen cycle. *In* "Nitrogen in Agricultural Soils" (F. J. Stevenson, ed.), pp. 721–772. Am. Soc. Agron., Madison, Wisconsin.

Tanji, K. K., and Mehran, M. (1979). Conceptual and dynamic models for nitrogen in irrigated croplands. *In* "Nitrate in Effluents from Irrigated Lands" (P. F. Pratt, ed.), pp. 555–646. Nat. Sci. Found., University of California, Riverside.

Taylor, R. M., and McKenzie, R. M. (1966). The association of trace elements with manganese minerals in Australian soils. *Aust. J. Soil Res.* **4,** 29–39.

Thomas, G. W. (1963). Kinetics of chloride desorption from soils. *J. Agric. Food Chem.* **110,** 201–203.

Thomas, G. W. (1977). Historical developments in soil chemistry: Ion exchange. *Soil Sci. Soc. Am. J.* **41,** 230–238.

Thompson, H. S. (1850). On the absorbent power of soils. *J. R. Agric. Soc. Engl.* **11,** 68–74.

Tien, C. T. (1987). Chemical reactions between some heavy metal ions and sludge particulate. Ph.D. Dissertation, University of Delaware, Newark.

Tillotson, W. R., Robbins, C. W., Wagenet, R. J., and Hanks, R. J. (1980). Soil water, solute and plant growth simulation. *Bull.—Utah Agric. Exp. Stn.* **502.**

Tripathi, P. .S. M., Tripathi, R., and Prasad, B. B. (1975). Radiotracer studies of adsorption of sulfate ions on ignited alumina. *Proc. Indian Natl. Sci. Acad., Part B* **41,** 156–159.

Ungarish, M., and Aharoni, C. (1981). Kinetics of chemisorption: Deducing kinetic laws from experimental data. *J. Chem. Soc., Faraday Trans.,* **1** pp. 975–979.

van Genuchten, M. T., Davidson, J. M., and Wierenga, P. J. (1974). An evaluation of kinetic and equilibrium equations for the prediction of pesticide movement through porous media. *Soil Sci. Soc. Am. Proc.* **38,** 29–35.

van Ophen, H. (1977). "An Introduction to Clay Colloid Chemistry." Wiley, New York.

van Riemsdijk, W. H. (1970). Reaction mechanisms of phosphate with $Al(OH)_3$ and a sandy soil. Ph.D Thesis, Agricultural University, Wageningen, Netherlands.

van Riemsdijk, W. H., and de Haan, F. A. M. (1981). Reaction of orthophosphate with a sandy soil at constant supersaturation. *Soil Sci. Soc. Am. J.* **45,** 261–266.

Velbel, M. A. (1985). Geochemical mass balances and weathering rates in forested watersheds of the Southern Blue Ridge. *Am. J. Sci.* **285,** 904–930.

Velbel, M. A. (1986). Influence of surface area, surface characteristics, and solution composition on feldspar weathering rates. *ACS Symp. Ser.* **323,** 615–635.

Vig, A. C., Biswas, C. R., and Sinha, M. K. (1979). Kinetics of phosphorus desorption and diffusion in differentially P fertilized soils. *Bull.—Indian Soc. Soil Sci.* **12,** 279.

Walker, A. (1976a). Simulation of herbicide persistence in soil. I. Simazine and prometryne. *Pestic. Sci.* **7,** 41–49.

Walker, A. (1976b). Simulation of herbicide persistence in soil. II. Simazine and linuron in long-term experiments. *Pestic. Sci.* **7,** 50–58.

Way, J. T. (1850). On the power of soils to absorb manure. *J. R. Agric. Soc. Engl.* **11,** 313–379.

Weber, J. B. (1972). Interaction of organic pesticides with particulate matter in aquatic and soil systems. *Adv. Chem. Ser.* **111,** 55–120.

Weber, W. J., Jr. (1984). Evolution of a technology. *J. Environ. Eng. Div. (Am. Soc. Civ. Eng.)* **110,** 899–917.

Weber, W. J., Jr., and Gould, J. P. (1966). Sorption of organic pesticides from aqueous solution. *Adv. Chem. Ser.* **60,** 280–305.

Weber, J. B., Weed, S. B., and Ward, T. M. (1969). Adsorption of s-Triazines by soil organic matter. *Weed Sci.* **17,** 417–421.

Wendt, H. (1966). Chemische Relaxation-suntersuchungen mit Hilfe des, Schallrohres im Frequenzbereich von 1 bis 30 kHZ. *Ber. Bunsenges. Phys. Chem.* **70,** 556–562.

Wolfrum, C., Bunzl, K., Dickel, G., and Ertl, G. (1983). Kinetics of isotopic and ion exchange in ternary mixtures of cation exchangers. *Z. Phys. Chem.* **135,** 185–199.

Wollast, R. (1967). Kinetics of the alteration of K-feldspar in buffered solutions at low temperature. *Geochim. Cosmochim. Acta* **31,** 635–648.

Wollast, R., and Chou, L. (1985). Kinetic study of the dissolution of albite with a continuous flow-through fluidized bed reactor. *In* "The Chemistry of Weathering" (J. I. Drever, ed.), pp. 75–96. Reidel Publ., Dordrecht, The Netherlands.

Wu, S., and Gschwend, P. M. (1986). Sorption kinetics of hydrophobic organic compounds to natural sediments and soils. *Environ. Sci. Technol.* **20,** 717–725.

Wyn-Jones, E., ed. (1975). "Chemical and Biological Applications of Relaxation Spectrometry." Reidel Publ., Dordrecht, The Netherlands.

Yan, F., and Xue, J. (1987). The desorption kinetics of surface adsorbed potassium in an external electric field. *Curr. Prog. Soil Res. People's Republic of China* pp. 45–52.

Yasunaga, T., and Ikeda, L. (1986). Adsorption–desorption kinetics at the metal-oxide–solution interface studied by relaxation methods. *ACS Symp. Ser.* **323,** 230–253.

Zasoski, R. G., and Burau, R. G. (1978). A technique for studying the kinetics of adsorption in suspensions. *Soil Sci. Soc. Am. J.* **42,** 372–374.

Zenz, F. A., and Othmer, D. F. (1960). "Fluidization and Fluid–Particle Systems." Reinhold, New York.

Zutic, V., and Stumm, W. (1984). Effect of organic acids and fluoride on the dissolution kinetics of hydrous alumina. A model study using the rotating disc electrode. *Geochim. Cosmochim. Acta* **48,** 1493–1503.

Index